"By exploring the intellectual and sociocultural conditions for doing quantum physics in 1920s colonial India from a non-Eurocentric angle, this book fills a void in the history of physics literature. Banerjee studies three important Indian physicists and identifies them as *bhadraloks*, a sort of Indian *Bildungsbürger*. Little more than their names and key contributions were known so far. This changes with this book, making it an important reference not just for historians of science, but for anyone interested in colonial history."

Dr. Christian Joas - Director, Niels Bohr Archive, Copenhagen

"Everybody familiarized with physics possibly have once asked how the Bose-Einstein statistics and the Raman effect were produced in colonial India in the early part of the 20th century. Banerjee's remarkable book mobilizes the cultural history and the idea of "*bhadralok* physics" to examine the group of intellectuals who looked for scientific goals as part of their social and cultural identities. The book fulfills an overdue gap in the history of physics."

Professor Olival Friere, Universidade Federal da Bahia, Brazil

"Banerjee grapples skillfully with a crucial question – how was early 20th century India able to develop advanced physics research while under colonial rule? The case studies of Bose, Raman, and Saha document a novel system of indigenous knowledge production - "*bhadralok* physics." This unprecedented book takes seriously the technical physics as well as the cultural diversity of India. It powerfully demonstrates how a postcolonial analysis can reveal an entirely new chapter in the history of modern physics."

Professor Matthew Stanley, New York University, USA

"In *The Making of Modern Physics in Colonial India*, Banerjee addresses a sadly neglected issue in the history of modern science and provides insightful answers to some central questions about the development of science. He shows how Indian scientists contributed major discoveries in the development of quantum physics, the most esoteric and novel branch of science at the time. He argues that this groundbreaking work was a part of the development of Indian nationalism and was very much a home grown phenomenon as major thinkers emerged from the *bhadralok* class of middle-class intellectuals and the movement for Indian self-rule. This engaging and well-written book is at once an important contribution to our understanding of scientific development in colonial and post-colonial societies and to our understanding of the development of quantum mechanics."

Professor Daniel Kennefick, Department of Physics,
University of Arkansas, USA

"This monograph offers a new perspective on the history of physics and modernity in early twentieth-century India. Transcending earlier approaches to global science as hybridity or the interaction of center/periphery or universal/local, Banerjee argues that early Indian quantum physicists reveal features of what he calls "cosmopolitan nationalism" or the melding of traditional Indian culture, British cultural traits and transnational ideas. This book situates the careers of three India-born and -educated physicists–Satyendranath Bose (1894-1974), Chandrasekhara Venkata Raman (1888-1970), Meghnad Saha (1893-1956)— within the Bengali middle-class intelligentsia known as the *Bhadraloks*. Living in a colonial situation, these physicists sought international collaborations outside the British empire and turned, in particular, to Germany. Banerjee's history of *bhadralok* physics is a fascinating study of colonialism and decolonialism, Indian nationalism and modernity, cosmopolitanism, and dynamics of class, caste and social manners. This book will engage a wide range of readers, especially those interested in science studies on a global scale."

Professor Richard Kremer, Department of History,
Dartmouth College, USA

"This is a fascinating and much needed account of the remarkable rise of physics in colonial India from humble beginnings to world-class status in quantum physics during the first half of the 20th century. Drawing upon his expertise in Indian and international scientific history and culture, Somaditya Banerjee provides an engaging story of broad interest and insight."

Professor David Cassidy, Hofstra University, USA

"This is a riveting analysis of how imperial science intersected with indigenous knowledge through the works of three twentieth-century scientists, S.N. Bose, C.V. Raman, and Meghnad Saha. It argues that their globally acclaimed interventions in relativity and quantum physics helped produce a radical variety of cosmopolitan nationalism. Somaditya Banerjee's engaging discussion of 'bhadralok physics' emerges as an essential reading on the intellectual lineages of colonial and Asian modernities, intersectional knowledge production, local and global histories of science and technology."

Professor Jayeeta Sharma, History Department,
University of Toronto, Canada

"How did quantum physics research start in India, while it was still under the British colonial rule? Somaditya Banerjee's book answers this question by studying the cases of three Indian physicists, Satyendranath Bose, Chandrasekhara Venkata Raman, and Meghnad Saha, whom Banerjee considers as "*bhadralok* scientists*." It is a fascinating read that explores physics and its socio-cultural context in India and expands the scope of the history of quantum physics beyond Europe and North America."

Professor Kenji Ito, SOKENDAI (The Graduate University for Advanced Studies), Japan

"Somaditya Banerjee's *The Making of Modern Physics in Colonial India* is a case study of three Indian physicists Meghnad Saha, C. V. Raman and Satyendranath Bose. All three of them made foundational contributions to the then nascent field of quantum mechanics. Their physics is now standard text book material. However, with his training in both physics and history, Banerjee brings a fascinating and new perspective to their story. Somaditya's unique contribution is the careful study of the socio-political and cultural milieu which played a key role in shaping Saha, Raman and Bose's science. This detailed and careful work is a very important contribution to the history of science in India, and should be of interest to scientists, social scientists, historians as well as the lay public."

Professor Jayaram Chengalur, National Centre for Radio Astrophysics (Tata Institute of Fundamental Research), Pune, India

The Making of Modern Physics in Colonial India

This monograph offers a cultural history of the development of physics in India during the first half of the twentieth century, focusing on Indian physicists Satyendranath Bose (1894–1974), Chandrasekhara Venkata Raman (1888–1970) and Meghnad Saha (1893–1956). The analytical category *"bhadralok* physics" is introduced to explore how it became possible for a highly successful brand of modern science to develop in a country that was still under colonial domination. The term *bhadralok* refers to the then emerging group of native intelligentsia, who were identified by academic pursuits and manners. Exploring the forms of life of this social group allows a better understanding of the specific character of Indian modernity that, as exemplified by the work of *bhadralok* physicists, combined modern science with indigenous knowledge in an original program of scientific research.

The three scientists achieved the most significant scientific successes in the new revolutionary field of quantum physics, with such internationally recognized accomplishments as the Saha ionization equation (1921), the famous Bose–Einstein statistics (1924), and the Raman Effect (1928), the latter discovery having led to the first ever Nobel Prize awarded to a scientist from Asia. This book analyzes the responses by Indian scientists to the radical concept of the light quantum, and their further development of this approach outside the purview of European authorities. The outlook of *bhadralok* physicists is characterized here as "cosmopolitan nationalism," which allows us to analyze how the group pursued modern science in conjunction with, and as an instrument of, Indian national liberation.

Somaditya Banerjee is an Assistant Professor in the Department of History and Philosophy at Austin Peay State University, Tennessee, USA. He is the Faculty Advisor for Phi Alpha Theta (which won the best chapter award in 2019 for the 11th consecutive time), History Club and the newly created India Club at Austin Peay and currently teaches Early and Modern World History, Historical Methods, Modern South Asia, Mughal India and History of Science & Technology.

At Austin Peay, Banerjee is also a member of the Faculty Senate and a member of the Hispanic Cultural Center Advisory Committee and the Adjunct Committee. Find him on Twitter @Soma_Band2020 or at https://www.apsu.edu/history-and-philosophy/faculty/banerjee.php

Empires in Perspective

Series Editor: Jayeeta Sharma, University of Toronto

This important series examines a diverse range of imperial histories from the early modern period to the twentieth century. Drawing on works of political, social, economic and cultural history, the history of science and political theory, the series encourages methodological pluralism and does not impose any particular conception of historical scholarship. While focused on particular aspects of empire, works published also seek to address wider questions on the study of imperial history.

The Making of Modern Physics in Colonial India

Somaditya Banerjee

Routledge
Taylor & Francis Group

LONDON AND NEW YORK

First published 2020
by Routledge
2 Park Square, Milton Park, Abingdon, Oxon OX14 4RN

and by Routledge
52 Vanderbilt Avenue, New York, NY 10017

Routledge is an imprint of the Taylor & Francis Group, an informa business

British Library Cataloguing-in-Publication Data
A catalogue record for this book is available from the British Library

Library of Congress Cataloging-in-Publication Data
A catalog record has been requested for this book

ISBN: 978-1-472-46553-5 (hbk)
ISBN: 978-0-367-49496-4 (pbk)
ISBN: 978-1-315-55579-9 (ebk)

Typeset in Times New Roman
by Deanta Global Publishing Services, Chennai, India

Dedicated to my mother and father—Dr. Roma Banerjee and Partha Banerjee

Contents

Figures

Preface and acknowledgments

When I was in physics graduate school at the University of Arkansas, I attended a physics colloquium on Albert Einstein and the history of gravitational waves by Daniel Kennefick. Dan's talk at the Paul Sharrah Lecture Hall was a life-changing event, so much so that I decided to switch careers from physics to history. My sincere thanks to Dan for introducing me to the very exciting field of history of science. When I arrived at the University of Minnesota, to do graduate work in the history of science and technology, Michel Janssen helped me think about the history of quantum physics and Sally Gregory Kohlstedt introduced me to the history of science and technology in various cultural settings. In the History Department at the University of Minnesota, I thank Patricia Lorcin and Ajay Skaria who introduced me to postcolonial theory and South Asian history. My sincere thanks to Alexei Kojevnikov at the Department of History at The University of British Columbia (UBC) for his incredible intellectual rigor, brilliant scholarship, and extraordinary generosity. At the UBC History department, I also thank John Roosa, who helped me think through South Asian history, and the most incredible Robert Brain, along with Harjot Oberoi, Paul Evans, Devendra Prakash Goel, CISAR and Allan/Kathy Abraham.

In North America and Europe, I thank David Cassidy, Rajinder Singh, Richard Staley, Matthew Stanley, Abha Sur, John Stachel, Lewis Pyenson, Asif Siddiqui, Antonia Moon, Roopen Majithia, Subrata Dasgupta, Joe Martin, Peter Pesic, Greg Good, Charles Day, American Institute of Physics, Niels Bohr Archive, Deepanwita Dasgupta, Angela Creager, Erika Milam, Suman Seth, Peter Galison, Richard Kremer, Ernie Hamm, Gustave Lester, Marco and Melinda Deyasi, Lori Celaya, Gordon McOuat, Robert Anderson, and Theodore Arabatzis. At Fayetteville, Arkansas, I'm thankful to Surendra Singh, Reeta Vyas, Arnabdyuti Mitra, Dileep Karanth, and the Rybas family (Pam, Ray, Adam, Ryan), and Brian Tessaro.

I thank all the archivists especially Pramod Mehra at National Archives and Nehru Memorial and Museum Library at Teen Murti House at New Delhi, Calcutta Mathematical Society, Saha Institute of Nuclear Physics and A.N. Sekhar Iyengar, Prof. Samit Ray at Satyen Bose National Centre for Basic Sciences, Bose Institute and T.P. Sinha, Chittabrata Palit, Institute of Historical Studies, Kolkata, National Library, Indian Association for the Cultivation of Science, Calcutta University, Jadavpur University, Indian Institute of Science, Raman

Research Institute (RRI-Digital Repository), Raju Varghese (photographer at RRI), Dr. Meera B.M., Indian Academy of Science, Delhi University Libraries and Archives, R.C. Yadav at Indian Institute of Chemical Biology, Amar Roy, Dhruv Raina, Shiv Viswanathan, State Archives, Kolkata, Intelligence Bureau, Kolkata, Enakshi Chatterjee, Subrata Dasgupta, Subhash Kak, Sarat Book House, Ambar Dey, St. Xavier's College, Presidency College, Richard Dickerson the curator of the Jagdish Mehra Special Collections at the University of Houston Library, Collected Papers of Albert Einstein at Caltech, Partha Ghose, Anadi Das, Satyendranath Bose's grandson Falguni Sarkar, Indian Science News Association, Bangiya Bigyan Parishad, Indian Statistical Institute, Asiatic Society Kolkata, Science & Culture, and Suprakash Roy. I also thank Vinod Kumar Rastogi, NRI Welfare Society, Deepak Singh, Hannah, Gauhar Nawab, Gopesh and Suparna Saha, Ajoy Ghatak, DCV Mallik, JIIT, Deepak and Rajani Dhingra, Anirban Pathak, Peter Minorsky, Baisakhi Bandyopadhyay, Rajeev Pathak and Pramod Joag at Pune University, Jayaram Chengalur and Rajaram Nityananda at NCRA.

Portions of my book have previously appeared in a few articles, and I am grateful to the editor and publisher of each journal for granting permission to use the materials here. Portions of Chapter 4 was published as "C.V. Raman and Colonial Physics: Acoustics and the Quantum", *Physics in Perspective* 16 (2), 146–178. Portions of Chapter 2 and Chapter 3 were published as "Transnational Quantum: Quantum Physics in India Through the Lens of Satyendranth Bose", *Physics in Perspective* 18 (2), 157–181. A portion of Chapter 5 was published as "Meghnad Saha: Physicist and Nationalist", *Physics Today* 69 (8), 38–44.

I thank the History and Philosophy department at Austin Peay State University (APSU). I don't think I have ever been in a more supportive and nurturing department than here at APSU. I thank Cameron Sutt, David Rands, Greg Zieren, David Snyder, Kevin Tanner, George Pesely, Dean Barry Jones, Minoa Uffelman, Ken Faber, Debbie Shearon, Nicole Wood, Greg Hammond, George Pesely, Christos Frentzos, Antonio and Amy Thompson, John Steinberg, Mark Michael, Jordy Rochleau, Dzavid Dzanic, Michele Butt, Mickey Wadia, Karen Meisch, Allyn Smith, Isaac Sitienei, Allan Chaparadza, Sergei Markov, Yoshio and Claudia Koyama, Felix G. Woodward Library, Joe Weber, my GTA Chesley Thigpen, Jessica Blake, and all my students at APSU. At Routledge, I greatly appreciate Rob Langham's support for the project and for steering the manuscript through the commissioning and editorial process. I thank the anonymous reviewers who helped me improve the manuscript considerably. Thank you very much Jayeeta Sharma, Tanushree Baijal and Rennie Alphonsa for all your help and making me stay on course. Any mistakes that remain are mine alone.

Finally, I thank my dearest mom and dad for cheerfully putting up with their only child living in distant North America for the last eighteen years. Without the exceptional support of my parents, it would be impossible for me to pursue my education in India and North America and it is to them I dedicate this monograph.

List of Bengali and Sanskrit words

Bhadralok:	Well-mannered, educated individual
Bhadramahila:	Female analogue of *Bhadralok*
Bharatiya:	Indian
Calcutta:	Called Kolkata in present day
Guru:	Revered Master or Teacher
Guru-Shisya:	Master-Student.
Jati:	Nation
Jatiyatabaad:	Nationalism
Madhyabitta:	Middle-income group
Pathsala:	School
Shesher Kabita:	The last poem (Title of a novel by Rabindranath Tagore)
Shishya:	Pupil or student
Swadeshi:	Of one's own country
Vande Mataram:	Hail to the Motherland
Vigyan/Vijnan:	Science
Visvajaneen:	Cosmopolitan
Visvajaneenata:	Cosmopolitanism

Abbreviations

AU: Allahabad University
BAAS: British Association for the Advancement of Science
CM: Classical Mechanics
CPAE: Collected Papers of Albert Einstein
CU: Calcutta University
CVR: Chandrasekhara Venkata Raman
DMB: Debendra Mohan Bose or Debendra Mohan or D.M. Bose
DU: Delhi University
IACS: Indian Association for the Cultivation of Science
IISC: Indian Institute of Science
INC: Indian National Congress
JCB: Jagdish Chandra Bose
JU: Jadavpur University
KSK: Kariamanikam Srinivasa Krishnan
MNS: Meghnad Saha
PCB: Prafulla Chandra Mahalanobis
PCR: Prafulla Chandra Ray
QM: Quantum Mechanics
RRI: Raman Research Institute
SINP: Saha Institute of Nuclear Physics
SNB: Satyendra Nath Bose
SNBCS: S.N. Bose National Centre for Basic Sciences
SSHRC: Social Sciences and Humanities Research Council (Canada)
UCS: University College Science (Calcutta University)

1 Introduction

Writing a history of modern science in South Asia

In 2012, physicists working at the European Organization for Nuclear Research with the world's biggest particle accelerator—the large hadron collider—made a watershed announcement.[1] They claimed to have experimentally observed signatures of the long-sought-after particle, the Higgs Boson, the discovery of which provided the decisive confirmation of the fundamental Standard Model in high-energy physics. The British physicist, Peter Higgs, who worked at the University of Edinburgh, had theoretically postulated the existence of the Higgs Boson in 1964. Overnight, the 2012 discovery made Higgs a major celebrity and the focus of much public attention, reminding of the way Arthur Eddington's observation of the 1919 solar eclipse had made Albert Einstein world famous by confirming the general theory of relativity.[2]

While the international scientific community celebrated a major victory for physics, an important aspect of the discovery went unrecognized. The new particle belonged to the fundamental class of bosons, named after the Indian physicist Satyendranath Bose (1894–1974). A modest college professor working in colonial India, Bose discovered in 1924 the special type of statistics that characterizes bosons as quantum particles. His proposal was supported, popularized, and further developed by Einstein. But for the rest of his life Bose remained in relative obscurity, teaching in India, and staying far away from major centers of physics. Even professional physicists of today, for whom bosons are a textbook concept that is constantly used in teaching and research, are typically aware of Bose's discovery, but know relatively little about its author, the context, and the circumstances of how it was made.

Interestingly, Bose was not the only Indian scientist of the early twentieth century who, while working in a distant country that was still under colonial domination, managed to make a breakthrough contribution to the emerging new field of quantum physics, and thus influenced the development of fundamental science in the European metropole. Such an unusual phenomenon, apparently without a historical analog, deserves a special reflection and investigation that is undertaken in this monograph from a multidisciplinary point of view, incorporating the social and cultural history of science, postcolonial theory, and the history of South Asia.

Overview

What constellation of circumstances allowed the development of an original and successful research program in modern physics in early twentieth-century India, a colonized country with limited financial resources, devoid of a fully institutionalized research, and with uncertain career trajectories for aspiring scientists? Yet, despite such hindrances, and in contrast to many other European colonies of the time, Indian scientists achieved their most important successes in one of the most sophisticated and revolutionary and cutting-edge fields—quantum physics. To answer this question thoroughly, we would need to conduct a historical investigation at several levels: the social context, the analysis of scientific works and intellectual influences, and biographical case studies.

This monograph first describes the social group to which those scientists belonged—the *bhadraloks*, or a new type of intelligentsia that developed under the special conditions of colonial power in India. *Bhadraloks* as a group were distinct from both the European officials as well as the traditional Indian intellectuals. They were natives of India who received European-style education and training, primarily for the purpose of assisting and working in the colonial administration. Yet many of their representatives defied or complicated that colonial agenda by turning into major promoters of the emerging Indian nationalism and the national independence movement. Bearing an ambivalent relation to the colonizers' heritage, the *bhadraloks* became the chief harbingers of the specifically Indian drive toward modernity that placed a particular importance and hope on science. Many of the *bhadraloks* pursued modern science in conjunction with, and as an instrument for achieving, independence from British rule, granting a significant role to science in the emergent Indian nationalism. In the process, some of them developed versions of science that sought meaningful connections between the modern, twentieth-century European scientific outlook, and the indigenous knowledge of India.

To illustrate and explore this distinctive historical phenomenon, this book analyzes in detail three individual cases: that of Satyendranath Bose, Chandrasekhara Venkata Raman, and Meghnad Saha. All three of them received their training and scientific education in India, but also used opportunities to travel and develop contacts with colleagues among European scientists. They all made important contributions to the revolutionary field of quantum physics, and achieved their greatest scientific successes in the 1920–30s. The notable ones are the Bose–Einstein statistics, the Saha equation, and the Raman Effect, which earned the Nobel Prize in 1930, and it was the first Nobel awarded to an Asian scientist. It will be argued that, for all three of them, the *bhadralok* identity is the key for understanding their lives and main accomplishments as scientists. At the same time, it is to be noted that they came from very different strata, traditional castes, and regions of India. Raman was a member of the upper social class and caste, Bose belonged to the middle caste, and, quite remarkably, Saha came from the lowest Indian caste, but managed to overcome the very strong social and cultural prejudices associated with it. Their correspondingly different versions of *bhadralok* careers and mentality affected their somewhat diverging scientific programs and results, and each of them articulated somewhat different versions of the cultural nationalism of the *bhadraloks*.

As was typical of the intelligentsia in many countries, science and higher education represented a major vehicle for social mobility between otherwise often rigid and hierarchical traditional classes and groups. One can observe this process in the different paths Bose, Saha, and Raman took to becoming *bhadraloks*, as will be explored in more detail in the subsequent chapters of the monograph. Raman was born into an educated, upper-class Brahmin family in South India, de facto inheriting the *bhadralok* status straight from his family background. Bose came from the middle tier, was born as a *Kayastha* and had to go through several tiers of academia before he was offered a teaching position in 1917 at Calcutta University. At that point he could be identified as a *bhadralok* by his intellectual pursuits. Saha's trajectory was the most challenging one, as he was born in the lowest tier of the caste hierarchy—the *shudra*. Saha had to face a lot of discrimination in his youth, especially in college, because of his lower caste. In spite of such adverse conditions, Saha eventually rose to become a professor of physics at Calcutta University in 1919. From there on, he too came to be perceived as a *bhadralok* by his intellectual peers and colleagues. To members of the Indian society, he appeared as an educated, civil, and well-mannered individual, possessing all of the attributes required for being a *bhadralok*. This was certainly not an easy task in the somewhat rigid, colonial Indian society of the early twentieth century. But Saha's search for an Indian modernity and the achievement of his *bhadralok* status made his accomplishments in science and his nation-building endeavors stand out.

The *bhadralok* careers of Bose, Raman, and Saha reveal some characteristic dilemmas between enjoying certain privileges and serving the colonial administration on one hand, and harboring nationalist aspirations on the other. They got their scientific training not in the metropole, but in India proper, and could be regarded as the first generation of indigenously trained modern scientists. They received financial support from the colonial government, their Indian mentors, and local philanthropists such as Sir Ashutosh Mukherjee. They approached modern science in a somewhat indigenous fashion, not by any means linked to the industrial hands-on approach exemplified by the Cambridge-British education. At the same time, they found a stimulating intellectual community and reference group overseas, mostly in Europe, as they communicated and collaborated with intellectuals, such as Albert Einstein, Niels Bohr, Arnold Sommerfeld, Alfred Fowler, and Walther Nernst. In the case of Bose, collaboration with Albert Einstein, who was a world luminary in the 1920s, was in keeping with his nationalist feelings, since it provided him an escape from the intellectual dependence on Britain. Raman's physics was essentially colonial in character, but with a fusion of indigenous and international traits. The academic trajectory of Saha, however, was much more uncertain than that of Bose or Raman, given his involvement with the Bengal Revolutionaries, along with a concomitant devotion to quantum physics. The subsequent chapters will explore in more detail and elaborate on the contradictions between colonial science and the specific project of Indian modernity.

Reflections of the social and cultural conditions of *bhadraloks*' scientific practices in colonial India can also be found in the ways Bose, Saha, and Raman responded to the revolutionary idea of the light quantum, which was the key scientific concept for their research. The chapter on Bose will argue that his education, the local cultural influences on his scientific beliefs, his anticolonial sentiments, and his fusion of nationalist aspirations with a cosmopolitan outlook were important for his acceptance of the quantum discontinuity of light. Saha's political radicalism converged with his wholehearted embrace of the radical concept of light quantum at a time when, unlike in the Indian colony, most scientists in the British metropole were still extremely skeptical of this subversive intellectual novelty that contradicted the well-established wave theory of light. The chapter on Raman will explore how his regional brand of nationalism, his fascination with Indian musical instruments and musical theory, his scientific work at the IACS and Calcutta University, and his dialogues with senior colleagues like Jagadish Chandra Bose, contributed to his biases towards the wave theory of light, and only a gradual, reluctant acceptance of the quantum theory of scattering as an explanation of the experimental effect he discovered.

The methodology adopted in this book combines the analysis of the scientific works of Bose, Saha, and Raman with an investigation into the social and cultural milieu in which their science was produced. The interplay between science and culture thereby informs the reader that their science did not operate in a social vacuum but was very much contingent on the culture of the period. To get a better understanding of the modernity of Indian physics in the 1920s, and to fuel public interest for the period under study, one needs a finer understanding of not only the technical components of the *bhadralok* physicists, but also the intellectual climate, the zeitgeist of the colonial period in South Asia.

The investigation of the three case studies will allow me to introduce the concept of "*bhadralok* physics" as a description and analysis of how modern science was pursued, and successfully developed in late colonial India. The main scientific accomplishments by *bhadralok* physicists—the Bose–Einstein Statistics (1924), the Saha ionization equation (1921), and the Raman Effect (1928)—reflect the culturally specific ways in which scientific knowledge is produced in the conditions of a colony striving for modernity. The results of this monograph thus bear ramifications for the two, typically separate, academic fields—South Asian history, and History of Modern Science—which have as yet remained mostly disconnected historiographically, but have the potential to contribute productively to one another. By developing these ties, my monograph contributes to the emerging historiography of modern South Asian science.

Reconfiguring the term *bhadralok*

The analysis undertaken in this monograph crucially depends on the somewhat malleable concept of *bhadralok* whose applicability, meanings, and attributions have received an evolving treatment in historical and sociological literature. To understand the *bhadralok* identity, it will be important to discuss how scholars have viewed the *bhadraloks* in South Asian history; who and how one could become a *bhadralok*, and who were not identified as such even among the elite and the middle class – for example, members of the business community, who were financially respected, but were not necessarily well-mannered, educated, and did not contribute toward the nation the way Bose, Saha, Raman, and other *bhadraloks* did. In addition to the three cases discussed in this monograph, there were many other scientists, such as A.K. Ramanujan, Debendra Mohan Bose, or Prasanta Mahalanobis, who can also be identified as *bhadralok* scientists and subjected to a similar analysis in future.

Histories of the *bhadralok* have been primarily written within the discipline of South Asian history, usually by historians who worked on social histories of nineteenth-century Bengal, but rarely had connections with or familiarity with the history of science. As a South Asianist, Tithi Bhattacharya remarked, echoing an argument by Sumit Sarkar:

> In their own perception this was a 'middle class bhadralok world which situated itself below the aristocracy' but 'above the lesser folk' engaged in manual labor and distinct from the lower castes or Muslims. What distinguished them from both was education of a particular kind, so much so that in commonsensical terms the pronouncements about education became the sole criterion for defining the bhadralok.[3]

Bhattacharya went on to say, "although the idea of the bhadralok is a necessary link in any analyses of nineteenth-century thinking and behaving, it is difficult to define."[4]

For Bhattacharya, the *bhadralok* category is applicable for the social organization in nineteenth-century India, and is strongly associated with the middle class, while excluding lower castes or Muslims. Including *bhadralok* scientists into consideration, as this monograph does, demands certain revisions to such an understanding. The more complicated case of Meghnad Saha, for example, demonstrates that even a person who originated from a lower caste could, under certain conditions, gradually transcend his rigid caste status through the pursuit of science and education to eventually becoming a *bhadralok*.

Another influential South Asian historian, John McGuire, highlighted some of the nuances of this social category, pointing out that it was much more than merely a label for "respectable people."[5] He argued that there are two problems in defining the term *bhadralok*. He suggested that it was rather problematic to use the term exclusively for Hindus and that the term "cannot be seen as a fixed social group, but rather as the embodiment of changing sets of organic social relationships."[6] Since McGuire considered only the period from 1857 to 1885, my book builds on his study. With the rise of the Indian National Congress in 1885 and the Partition of Bengal in 1905, many Indian nationalists became identified as *bhadraloks*, while historians gradually reified *bhadralok* as a middle-class and not a lower-caste entity within the nation.

For example, another South Asian historian, Amit Kumar Gupta, in his monograph, *Crises and Creativities: Middle-Class Bhadralok in Bengal c. 1939–52*, stated that

> the bhadralok in Bengal formed essentially a socio-cultural category who had a distinct way of life, characterized by a certain standard of personal and familial refinement, a code of public and societal conduct, and a well-laid out system of values.[7]

Gupta continues to say that

> socially the category incorporated all members of the middle class (both the upper and the lower), excluding strictly those who performed manual labour of any kind, and those who were educationally handicapped, but included liberally the members of the rich.[8]

Gupta considers the possibility of a lower class to be included in the *bhadralok* category, provided there was no manual labor performed. Gupta does mention that *bhadraloks* were typically English educated and economically stable but "were dominated by the upper caste Bengali Hindus."[9] I disagree with him on the point about upper caste Bengali Hindus, because anyone with education, refinement, and manners could actually become a *bhadralok*.

Similarly, Swati Chattopadhyay another South Asianist remarked that:

> In nineteenth-century Bengali parlance the landed elite were referred to variously as *vishayi* (propertied), *dhani* (wealthy), *abhijata* (aristocrats), or *baramanush* (literally, "big" people), and the middling classes were referred to as *madhyabitta* (middle income) or *grihastha* (householder). They, along with the "*daridra athacha bhadra*" (poor yet respectable) constituted the respectable minority of the Bengali residents in the city – the *bhadralok*. Freedom from manual labor (for men) was the prime factor that designated these classes/caste as "respectable," a factor that distinguished them from the lower classes/castes or *chotolok*. The respectable stratum typically consisted of the higher castes of Hindu society.[10]

If we analyze this passage, we find that unlike Chattopadhyay, one can make a distinction between class and caste, the two systems of social stratification. Caste is a unique system of social stratification prevalent in India, where status is fixed by birth, whereas class allows mobility between strata. As there is no mention in her remark of the possibility of social mobility in Indian society so far as class distinctions are concerned, this monograph tackles this problem by examining the *bhadralok* physicists.

As *bhadralok* is a Bengali word that ascribes positive values to an individual who is polite, gentle, and well-mannered (*bhadra*), there is also the *abhadra* who do not qualify in the *bhadra* category. The *abhadra* are impolite, use foul language, and can belong to any class or caste. Thus, *abhadra* is the negation of *bhadra*, emphasizing the social distinction that transcended caste and class categories, or at least modified them. *Abhadra* can also be called *chhotolok*, or a lowly person. Just as a person can rise to the *bhadra* level, a *bhadra* person who is educated and well-mannered could, by his actions, become *abhadra*. This transition is powerful because once a person is de-classified as *abhadra*, one has to remain as an outcast in society. The *bhadra* status could be lost in a day through marriage, or if the family's reputation is maligned.

The three *bhadraloks* examined in this monograph all married as per family instructions, arranged within their social groups, to very young women. Women also had to belong to refined, or *bhadra*, families. Such refined and educated women were called *bhadramahila*, where *mahila* means female in Bengali. Saha's family had to overcome an old prejudice concerning the fact that they had once specialized in the brewing and distilling of alcohol, which was not a "clean" profession for a *bhadralok*. Some occupations, such as pathologists, printers, and pharmacists, took years to move up the scale to "clean" occupations. One strategy was to undergo suitable marriages with *bhadra* women; therefore, this high status was not simply restricted to male roles and male occupations. If anything, this status was just as much about women—potential wives and daughters-in-law—and their relations to *bhadralok* men. The major capital outlay for a middle-class male was for the marriage of his daughter(s), so this investment required extreme care. The opinions of the wife and her female relatives were decisive in the matter of marriage. This was one sphere in which women were more important than men within the *bhadralok* culture.[11] While the role of gender within *bhadralok* is an important issue, existing scholarship has also focused on the modernity of these Indian intellectuals who pursued science.

Dhruv Raina and S. Irfan Habib argued that "The Bengali Bhadralok class was a Western-educated modern elite who had been socialized through the colonial education system into 'colonial values'"[12] While this statement might be valid for Indian scientists who were Western educated like Jagadish Chandra Bose and Prafulla Chandra Ray, it cannot be applied to the first generation of indigenously trained scientists like Satyendranath Bose, C.V. Raman, and Meghnad Saha.

Raina and Habib echoed Mahendralal Sircar—the founder of the Indian Association for the Cultivation of Science (IACS)—when they further argued that

> for a colonial subject, the inauguration of the age of modernity is imbued with an inescapable ambiguity: it is an age of invasions and oppression; but, in addition, for those who empathize with the project of modernity, it is an age of advancement of the sciences.[13]

In their study, Raina and Habib focused on the social context of science without an actual engagement with the contents of scientific research. Thus, their analyses were based on a form of modernity which was not easy to explain. I think that such an externalist approach to social history in India, as taken by Nehruvian[14] historians, reflects the academic field of the history of science, which is still in a formative stage today. This monograph, however, builds on the work of Raina and Habib and approaches the period under study from a dual externalist and internalist approach, so that one has a fruitful interaction between South Asian culture and the contents of science influenced by the *bhadralok* culture. This, however, begs the question of who these intellectuals were.

The *bhadraloks* were trans-class, trans-caste individuals who were well-mannered and polite. An individual born in the lowest *shudra* caste could move up the social ladder and achieve the status of *bhadralok* through the acquisition of higher education. So, the term *bhadralok* has a sociological implication as it signifies a new status in Indian society. For example, Meghnad Saha was one of the greatest *bhadralok* scientists India has ever produced, because he worked hard toward reaching the status of *bhadralok* while coming from the lowest caste *shudra*.[15] Therefore, this conceptualization of Saha rising in Indian society from a *shudra* to a *bhadralok* through his science is new, as presented in this monograph. The approach of these colonial intellectuals was unique with regard to their approach to science.

Physicists among the *bhadraloks* blended Western culture and physics with Indian tradition to create what I call a unique "cosmopolitan nationalism" that— being somewhat similar to the German "mandarins"[16]—served at once to foster the national culture, to divert support from the political authorities of the time, and to promote its adherents into the upper social and scientific strata. Subsequently, the similarity between *bhadraloks* in India and Wilhelmian[17] academic scientists is striking.

Russell McCormmach, in his seminal article "Academic Scientists in Wilhelmine Germany," developed ideas of the German "mandarin" culture in the famous work of Fritz Ringer: *The Decline of the German Mandarins*.[18] McCormmach, echoing Ringer, argued that

> the cultural justification rested on the self-appointment of Wilhelmian academic scientists to the class of culture bearers. 'Culture-bearer' (*Kulturträger*) was a value-laden term denoting those who were considered well educated and qualified to judge matters affecting the quality of culture.[19]

Just as "German scientists had long placed their scientific ideology in the service of their greatest political cause, the unification of Germany,"[20] the Indian *bhadraloks* I examine in this monograph steered their scientific pursuits in direct and indirect ways toward Indian independence and decolonization.

Moreover, the concept of *Kulturträger* is important in this context because it carried a somewhat anti-Western, especially anti-British, connotation, implying that Germans were a people of culture (*Kultur*), whereas, the British were people of civilization, i.e., materialist values. German scientists, like Hermann Helmholtz, who adhered to this ideal, not only pursued culture avidly, but also signaled it in their general bearing and in the myriad ways they executed science. For example, Hermann Helmholtz in *Erhaltung der Kraft* (1847) accepted this *Bürgerliche Intelligenz* as science's primary task.[21] Thus, the Prussian educated members of the bourgeoisie—the *Bildungsbürgertum*—and the Indian *bhad-ralok*'s rationale for executing science were humanistic, and a narrative of Indian modernity would remain incomplete without fleshing out these transnational connections. Furthermore, Gerald Holton has written about Einstein from this perspective, identifying him as a German *Kulturträger*.[22] This monograph builds on this existing innovative scholarship applying it to the South Asian context.

This information is relevant because C.V. Raman—another *bhadralok* scientist examined in this monograph—considered Helmholtz to be his scientific *guru*. And Raman's investigations of Indian musical instruments were very similar to what Helmholtz exemplified in his *Sensations of Tone*, in which he showed how Western musical theory was elaborated in and through the character of Western musical instruments, which had evolved just as the physical science of acoustics had developed. As a result, both Raman and Helmholtz were trying to indigenize both science and global musical theory. Hence the nineteenth-century *deutsche Kultur*, as seen through the *bildungsbürgertum*, and the twentieth-century *bhad-ralok* culture had a few similarities and differences.

As the subsequent chapters will show, Indian *bhadralok* scientists embraced German science as a means of getting away from the colonial Indo-British framework. German scientists, such as Arnold Sommerfeld and Albert Einstein, were in turn impressed by the culture and bearing of Indian scientists who added a dressing of credibility to their excellent scientific work. While Sommerfeld really liked Saha and Raman, and Bose–Einstein statistics is one of the triumphs of twentieth-century physics, this was not always the case especially with some German scientists, such as Richard Gans at Jena, who was very skeptical of Raman's experimental work. However, it is unclear whether *Bürgerliche Intelligenz*[23] (science as a cultural project) was the primary project of German physicists interacting with the *bhadralok* intellectuals. Such close entanglement between German and Indian physics leads us to examine the scholarship on the history of physics.

History of physics

The history of physics has long been plagued by debates between "internalists" and "externalists." Broadly speaking, internalists concerned themselves with the technical and conceptual development of physics while externalists were motivated by society, politics, and institutions. A classic work in the history of physics written by Paul Forman showed how this debate is actually not inappropriate for writing a cultural history of science.

Forman argued that indeterminism, or acausality, in quantum physics appeared because of Weimar culture.[24] The acausal description of events governing the dynamics and kinematics of the subatomic world came about as a purposeful adaptation by physicists and mathematicians to the hostile intellectual milieu in Weimar Germany. After the end of World War I that brought defeat and devastation to Germany, the political, cultural, and intellectual climate became irrational. The military defeat, financial uncertainty, and social crisis prompted many intellectuals to question the Enlightenment ideals of rationality and progress and inspired corresponding criticisms of science. The Forman thesis launched a heated discussion with polarized views among historians of science, but in the long run proved very influential in ushering new approaches to the history of science from a cultural standpoint. My approach in this monograph is also indebted to Paul Forman insofar as I examine the influence of the cultural milieu—both external and internal to science—in the making of modern physics in colonial India; only this monograph will consider both internal and external factors as important elements in the cultural history of South Asian science.

One important caveat for the cultural history of science is that the cultural values that are prevalent spatially and temporally exert influences on scientific research, including the content of science, as revealed in the history of quantum physics. Paul Forman's work was instrumental in the rise of new approaches to science studies, and the history of physics during the 1980s. A growing number of case studies involving various cultures and different fields of sciences have emphasized a now widespread understanding that science is produced and co-produced locally in particular cultural settings.[25] Despite these growing examples of cultural histories of science, there has been precious little work of culturally based studies of physics from the early twentieth century till the time Forman published his work. From the 1980s to the late 1990s the academic landscape started changing with Peter Galison's and Andrew Pickering's work on important cultural analyses of early twentieth-century physics, especially relativity theory and particle physics.[26]

However, there still remain some blind spots in the current literature on the history of quantum physics. The contributions of South Asian scientists are either absent or misunderstood due to lack of a close reading of their lives and works. For example, never has Forman's influential work on "Weimar Culture and Quantum Acausality" been extended to other "contact zones" like South Asia.[27] Despite very few recent works in this direction, "we know little about the social, political, and cultural influences on the content of scientific knowledge produced in India."[28] This shortcoming highlights the need for more cultural histories of physics within a colonial framework. My monograph is a beginning piece in that direction which builds on the existing scholarship on cultural histories of science and applies it to the Indian context.

A crucial aspect for the development of modern physics in India was the weakness of certain established traditions. For example, scientists' perception of the Maxwellian electrodynamical continuum was not as embedded in India as in Europe, especially Britain. India lacked a tradition of classical physics. This crucial absence played a key role in the easier acceptance of Einstein's light quanta after 1905, an acceptance which did not happen very smoothly in Europe. Consequently, the reception of quantum discontinuity was very different in India. More importantly, this specific field of history of Indian physics is still an unexplored territory for the social historian whose focus is the social and cultural underpinnings of science. It is also unexplored because some of the representations of Indian scientists, as perceived in Europe and North America, are not based upon a close textual analysis of their works. For example, Einstein's biographer Abraham Pais denotes the work of Satyendranath Bose as "serendipitous." My investigation of Bose's approach to quantum statistics will reveal the cultural factors on which Bose–Einstein statistics were contingent. This study will rely on the forthcoming analysis of several other classic works in the history of quantum physics.

Mara Beller has written about the history of quantum physics and its interpretations by important interlocutors, like Niels Bohr and Werner Heisenberg among others.[29] Beller used the phrase "quantum dialogue" as a lens to see through the maze of intellectual conversations, which were not always free from paradoxes and uncertain interpretations. However, as Beller argued, these dialogues helped assist in furthering the emergent field of quantum mechanics, and its dominant "Copenhagen interpretation."[30] Beller's dialogue-thesis is innovative and inspiring. It can, with some modifications, be applied to the colonial situation in which the Indian physicists worked like Satyendranath Bose's contributions in quantum statistics, C.V. Raman's work in light scattering, or Meghnad Saha's work in astrophysics.

Richard Staley analyzed the basic categories of classical and modern physics as historic constructs useful for periodization, which were first introduced by Max Planck during the 1911 Solvay conference.[31] The time around World War I is generally seen as a watershed that saw the collapse of the classical "world picture" and the ushering in of modern physics.[32] This monograph will explore the reactions to these events by the Indian intelligentsia, which did not actively participate in the debates until the end of World War I; and their response to the newly emergent quantum physics of the non-classical discontinuous theory of light and the modernity it entailed. It is also important to highlight that modern physics and modernity in physics are subtly different yet related entities. While modern physics refers to drastic changes in the way physicists conceptualized fundamental theories like quantum physics early in the twentieth century, the onset of modernity in Indian physics began when physicists started explaining novel phenomena using traditional classical explanations, for example, the endeavor of physicists in India to explain the Compton Effect using classical physics. The transitional passage to modernity continued with more non-classical phenomena, like the existence of spontaneous emission that Einstein tried to explain phenomenologically in 1917. But Indian scientists, as I examine here, gave a statistical explanation of spontaneous emission on which Paul Dirac later worked to produce a full-fledged quantum electrodynamics. This was the background to the modernity of modern physics.

The transition to modernity was complete when there was a switch from what was mechanical and visualizable (e.g., orbits), to a mathematically abstract, non-visualizable (e.g., transitions) and counterintuitive domain of matrices and non-commuting algebra, as seen in a formalism of quantum mechanics called matrix mechanics, which emerged in 1925.[33] The experimental verification of matrix mechanics was given by the Raman Effect in colonial India, discovered in 1928 by a *bhadralok* intellectual, C.V. Raman, and his cohort working in Calcutta. Hence, these *bhadralok* intellectuals need to be studied not only because they are important for discoveries in physics and its history, but also because several approaches in postcolonial theory have toiled hard to understand the development of modernity outside the purview of Europe.

Postcolonial theory

While theories of modernity, as formulated by historians, anthropologists, and, most recently, postcolonial theorists, have come to mean a wide variety of things, it is most useful to contextualize them historically, being sensitive to the various perspectives that exist in present scholarship.[34] A historian at NYU, Frederick Cooper, has given an insightful analysis of modernity using a four-fold definition in the context of colonialism. First, Cooper argued:

> modernity represents a powerful claim to singularity, which is central to the history of Western Europe and a goal which the colonized world aspires to acquire as a tool to break from the shambles of backwardness. Secondly, it might be an imperial construct which gives the ethical right to the West to impose its will on the colonies. Thirdly, the singularity and European nature of modernity will always make it an unattainable object by the non-European world, however close one may come. Fourthly, the nature of modernity can be espoused in different plural cultural, local and transnational forms and there exist multiple modernities and alternative modernities.[35]

Giving some agency to the non-Western world, this fourth category shows how non-European cultures could engender unique forms of representations and conditions of modernity. These are not mere mimicry of Western modernity but, in actuality, attempts to derive alternative techniques that are self-consciously distinct and independent of colonial connotations, as we will explore through our case-studies of *bhadralok* intellectuals. For example, postcolonial theorists like Partha Chatterjee objected to the notion of colonial India being a passive recipient of Western modernity, and being reduced to the role of "perpetual consumers of modernity."[36] Chatterjee's works showed that "the colonial intelligentsia was pondering over the issue of Indian nationalism in the light of a different modernity and made a distinction between 'our modernity' and 'their modernity'."[37] This scholarship made a splash originally with the publication of an important text in the late 1970s.

In 1978, with the publication of *Orientalism*, an influential work by the Palestinian-American literary theorist and public intellectual, Edward Said exerted a remarkable influence in South Asia,[38] especially on discourses about knowledge produced in the colonies and the various brands of nationalism beginning from the 1980s. Said's *Orientalism* by itself pays no serious attention to British Orientalism in South Asia, and mostly concerns itself with scholarship on the Middle East. The nature of the Saidian discourse initiated a manifest tradition, engendering an outpouring of specific writings on India defined as the Orient.

In the wake of Said's *Orientalism*, two evaluations have emerged amongst historians. The first, most notably Gyan Prakash, following Said's thesis, contends that the discourse of Orientalism was hegemonic as extended to South Asian intellectual history. The second evaluation, following Kapil Raj, claims that "colonized South Asians played a determinant role in a dialogical process through which 'colonial knowledge' was constructed."[39] But what about voices that were not included in nationalist narratives? What about subaltern[40] voices?

The study of South Asian history was meant to further develop with the ushering in of the Subaltern Studies Collective (SSC) in 1982, along with the displacement of Marxism as the dominant mode of theoretical framework amongst South Asian historians engaging in the relationship between the Western and non-Western worlds. The problem that a group of South Asian academic scholars, like Ranajit Guha and Gyanendra Pandey, addressed was how to write a "history from below," i.e., a history about the ordinary people at the grassroots level, the non-elites of India. Vivek Chibber argues that "while elite politics could be identified with the modern institutions built around the colonial state, the domain of the subaltern constituted a distinct arena which was different from the ruling elite with a manifestly important yet underrepresented area".[41] But what about the scientific non-elites, the non-dominant social groups whose career trajectories show that they had their own conception of the world which differed from the mainstream nationalists? While expanding the idea of the marginalized, the disenfranchised, or popularly, the term "subaltern," Antonio Gramsci remarked:

> In acquiring one's conception of the world one always belongs to a particular grouping which is that of all the social elements which share the same mode of thinking and acting ... When one's conception of the world is not critical and coherent but disjointed and episodic, one belongs simultaneously to a multiplicity of mass human groups.[42]

The Subaltern Studies Collective (SSC) in the early 1980s made a major impact on South Asian history especially by examining various intellectuals who were marginalized and hence could be seen as subaltern. However, this approach was never extended to Indian scientists who could possibly be conceptualized in this category. My monograph is not an intervention in using the Subaltern Studies framework, but is an inspiration on a motivational register to write about *bhadralok* intellectuals who were neither subaltern nor elite. This book espouses a new approach in Science Studies in South Asia by looking at Indian scientists who were not exactly elite or subaltern, and demonstrates how their conceptualization of the light quantum differed from the dominant notions of light that existed in Europe. Such an approach goes significantly beyond the common view about the fusion of separate cultural and knowledge traditions that is typically theorized using a hybrid model of the colonizer and the colonized.

One of the postcolonial/Subaltern theorist Homi Bhabha's insights looked at the Indo-British encounter in a binary-mode, as an interaction between two well-defined entities (hybridity).[43] Based on this assumption, the notion of hybridity was made popular, and the concept has acquired a widespread following in North America. Analyzing the colonial intelligentsia in early twentieth-century India shows the problematic nature of this notion of hybridity. Especially in the sciences, hybridity, (i.e., cross-fertilization between two traditions) does not capture the full problem of explaining how Indian scientists produced new knowledge. In fact, there were many elements in Indian science that were neither Indian nor British but belonged to the wider transnational community. While a broader framework is required to examine the making of modern Indian science, it may be pointed out that there was exchange of scientific knowledge between the local and the global. Knowledge was also indigenized, leading to the development of a distinctly modernized yet local form—amalgamation of tradition with modernity. One needs to analyze the rich scholarship on the history of science in India, and how science and nationalism have interacted within a power differential in a colonial landscape.

Science studies in India

Analysts of science and technology in India have examined the varieties of nationalism expressed by scientists of the pre-1950 period. However, a complete understanding of how scientists were influenced by nationalism, or even the wide spectrum of nationalist aspirations that were also internationalist and transnational while grounded in the local, is still developing. For example, John Lourdusamy has studied four individuals—the Indian homeopath Mahendralal Sircar, the philanthropist and educator Ashutosh Mukherjee, the chemist Prafulla Chandra Ray, and the physicist/plant physiologist Jagadish Chandra Bose. These individuals played different roles and interacted differently with the subjects of this monograph. So, it is important to mention them in the present context, and also to mention the problematic aspects of this scholarship.

Lourdusamy argued that Sircar, Mukherjee, Ray, and Jagadish Chandra Bose's "engagement with western science was not a nativist project of identifying an exclusive Indian science, but was a confident and positive engagement with a universal modern science".[44] Lourdusamy claimed that Sircar, a prominent practitioner of homeopathy in Calcutta and the founder of the Indian Association for the Cultivation of Science (1876), established the Institute to promote scientific research among Indians, a project that led to the emergence of nationalist movements.[45]

Furthermore, Lourdusamy argues that the well-known physicist cum plant physiologist, Jagadish Chandra Bose, "sought to infuse elements of Indian culture into western science from a conviction that science was a global heritage".[46] Noted chemist, Prafulla Chandra Ray, who contributed greatly to modern chemistry by discovering an amorphous form of a chemical compound—mercurous nitrite—in 1896, established the Bengal Chemical and Pharmaceutical Works (1893) and wrote the *History of Hindu Chemistry* (1902), which viewed Indian history in very discontinuous terms characterized by phases of order and external invasions. Ray made important scientific contributions to the metropolitan science community while also pondering over the low rate of literacy in colonial India. Lourdusamy's thesis, as Pratik Chakrabarti remarked, claimed that "the works of the Indian scientists were not non-conformist practices from mainstream modern science but were very much in keeping with universality."[47]

While Lourdusamy's detailed account of the lives of the scientists is informative, it is also very descriptive. It is not clear what Lourdusamy meant by universal nature of modern science as perceived by Indian scientists. His work fell largely within a framework highlighting the agency of Indian scientists in their selective adoption of Western science, and coupling this agency with nationalism.[48] It is unclear in Lourdusamy's narrative what role nationalism played for scientists in late-nineteenth- and early-twentieth-century India.

"Nationalist consciousness" should not necessarily be equated with nationalism. For example, the Indian National Congress (INC from here on), which was created in 1885, and its supporters and collaborators among scientists, argued in the early years that India had never been a nation. The nationalist movement gathered steam after the Partition of Bengal in 1905, and between 1915 and 1920, when INC began to organize mass protests under the purview of Mahatma Gandhi's leadership. Lourdusamy's work makes no mention of the renowned Indian nationalist thinker Satishchandra Mukherjee, who launched the Dawn Society to promote the idea of national education. With the formation of the Indian Science Congress Association in 1914, Indian scientists gained a wider platform on which to assemble and exchange ideas. These developments, which are missing in Lourdusamy's narrative, can be argued to have been important reasons for the development of a nationalist science.

Whether nationalism was a result of interaction with foreign knowledge systems is debatable, but science was one of the most important components of the Indo–British colonial encounter. It will also be clear from this monograph that knowledge was not transferred to India in a passive way. However, the ideology of science was reconfigured in the zeitgeist of India's culture. Science and nationalism were also closely enmeshed, especially with nationalists who thought beyond the nation, and whose intellect displayed a cross pollination of local, national, and global ideas. For example, Subrata Dasgupta, another scholar of science studies in India has given a more thorough analysis of Jagadish Chandra Bose (JCB from here on).[49]

Though Dasgupta distanced his account from nationalist historiography, he argued that for JCB and many of his contemporaries, "knowledge and glory were inextricably intertwined with the Indian past."[50] Dasgupta wrote that

> JCB's experiments initially in the electromagnetic theory and later with metals and plants were a vindication of his interpretation of ancient Indian wisdom and Vedic monism. JCB felt that what India needed was not a few individual scientists, but a rejuvenation of a long-lost treasure of its scientific knowledge to generate a whole institutional framework of scientific research.[51]

Although Dasgupta was aware that colonial relations, nationalist ideologies, and metaphysical commitments of Vedic monism played an important role for Jagadish Chandra Bose's creative work, he nonetheless adhered to a "rather strict separation between science and extrascience."[52] In this book, Dasgupta's analysis will be further extended to JCB's mentees—Bose, Raman, and Saha.

Insofar as the mentors of the scientists in this book are concerned, Prafulla Chandra Ray, the author of *The History of Hindu Chemistry*, had a *Weltanschauung* that was inspired by Indian history. His involvement in indigenous chemical research relied on looking back at the ancient Indian engagement with the chemical element mercury. Dasgupta argued that Ray was successful in forming a school of chemistry, while Jagadish Chandra Bose failed to create a school of physics. His argument was that Jagadish Chandra Bose failed because he was an upper class and upper caste scientist, while Ray achieved success in creating a school of chemistry because he was from a "lower" caste, as per the traditional Indian caste system.[53] Instead of a failure–success binary, I argue it would be instructive to delve deeper into the methodologies of the *bhadraloks* and the multidimensional nature of their science. Before that, it is important to examine the very recent scholarship on nuclear science in colonial India, because the actors of this book created a strong platform on which nuclear science launched itself in the 1940s.

Few recent studies in the past decade have discussed the key role played by science, particularly the quest for nuclear energy and its symbolic power. In 2010, an influential study by Robert Anderson gave a very detailed ethnographic history of scientists and scientific establishments, starting from 1920 till 1980, that played a significant role in integrating state-making with nationhood. Using the Actor Network Theory as methodology, the author has traced the "nucleus" of people who made the creation of the first atomic bomb possible in India in 1974, and the nucleus' relation to the nation. The nuclear program in South Asian historiography is used as a "focusing device to understand the scientific community."[54] The main focus of the study examined the Indian physicist Meghnad Saha, the chemist Shanti Bhatnagar, the physicist Homi Bhabha, and nationalist leaders like Mahatma Gandhi and Jawaharlal Nehru. Anderson argued that "that there would not have been a sustained atomic energy program in India without a co-evolving relationship between science and politics, which resulted in a larger community."[55]

The most innovative conclusion Anderson reached is that "there has been a creative tension running through the Indian scientific community between the idea of science as a movement and science as an institution."[56] The author described this tension as a process of *schizmogenesis*, the term coined and described by the anthropologist Gregory Bateson in *Naven* (1958), which is in essence "a process of separation and disconnection" at the levels of both rhetoric and action. For example, Saha Institute's emergence from Science College Calcutta and also Raman Research Institute's emergence from Indian Institute of Science can be seen as examples of the tension between science as a movement and science as an institution.[57]

While Anderson's brilliant account contributed richly to the complex interplay between science and culture in the Indian context, and extended Science and Technology Studies to non-Western cultures, my monograph approaches the problem of the origins of Indian science in the early twentieth century by the first generation of indigenously trained *bhadralok* scientists. These scientists created a platform from which, several decades later, modern physics in India became institutionalized in such a way as to tackle the nuclear and, to use Itty Abraham's phrase, "grew to love the bomb."

In contrast to Anderson's study, Itty Abraham's work on how India "grew to love the bomb" does not pay sufficient attention to the intricacies of science in India, or to the understanding of its scientific community. A historian of science, Deepak Kumar, commented on Itty Abraham's work:

> How did science figure in this debate? What constituted India's colonial heritage? In addressing these questions Abraham refers, rather uncritically, to Gyan Prakash's thesis on "Hindu" science and revivalism. Did scientists such as P. C. Ray and others try to establish Vedic Hinduism as the preeminent definition of Indian traditions? Definitely not. After a brief comment on "colonial science" in the following chapter, Abraham introduces Homi J. Bhabha, the father of India's nuclear program, as a "colonial scientist who brings to the fore the anxieties and ambivalences of metropolitan Western science." How Bhabha did so is not really made clear.[58]

Deepak Kumar's critique of Abraham is part of a larger problem of narratives on Indian science that make the Indian nuclear program synonymous with Indian science, and also engage predominantly with elite figures of Indian nationalism, such as Gandhi and Nehru. It seems as if one cannot write a history of India without hagiographies on figures like Bhabha, Gandhi, or Nehru. On this point, I appreciate the existing scholarship, but also differ from it.

The process of historical investigation for me is not restricted to a narrow engagement with the elite characters mentioned above, and a celebration of their careers, but with having to situate the investigation in an extensive horizon involving many individuals who were not necessarily elites, but could be conceptualized as *bhadraloks*, and have been overlooked in historical narratives.

While there is plenty of literature on the Indian nuclear program and the atomic bomb project, historians of South Asia seem to be "oblivious" of the role of science in general. What is more, this lacuna has happened because most South Asian historians focus on exclusively social histories, and typically omit the scientific content from their analyses. As Prakash Kumar correctly remarked in his 2012 monograph on indigo:

> But the study of science in South Asian historiography has so far evolved along two parallel tracks – works that cover colonial science and works that cover the social history of science in colonial South Asia. Their respective philosophical orientations and theoretical borrowings have led them in different directions and they have built their own respective momentums in isolation from one another. Thus, South Asia historians who study "science" fall into one group or the other. The partiality in favor of analysis in one or the other framework also accounts for the apparent chasm that separates the

study of science so far. This mutual obliviousness is unfortunate because each field has much to contribute to the other.[59]

Indeed, most South Asian historians have stayed away from deliberations of science; while those who study science, like Gyan Prakash and Kapil Raj, belong to an extreme "externalist" category that uses discourses around science, and the images of science as their primary modus operandi in creating narratives. While these issues surrounding discourse and images of science are important issues, they do not fully capture, for example, conditions of how colonialism, nationalism, cosmopolitanism, and local knowledge systems influenced the growth of the character of scientific knowledge. My monograph draws material from both history of science and South Asian history to produce a narrative that aims to resolve this "mutual obliviousness," thereby bridging the "chasm" that, according to Prakash Kumar, exists between these two fields.

A more balanced account of Science and Technology Studies in India, developed by digging deeper into the technical contents of the works by Indian scientists, will help substantiate claims made by South Asian historians who approach these issues using cultural history and postcolonial theories, and ascribe "difference" to South Asian scientists without a serious engagement with their research. For example, the Princeton historian Gyan Prakash has traced the genealogy of the culture of Western sciences in India in the nineteenth and twentieth centuries. Gyan Prakash remarked that "the insistent demand for a nation-state was an urge to establish a modernity of one's own, one that differed from Western modernity."[60] It remains, however, unclear from Prakash's narrative in what way Indian modernity was different from Western. Hence it is important to flesh out how science developed during the British rule, especially from 1876 on, which was the founding moment for the first indigenous institute of scientific learning—the Indian Association for the Cultivation of Science (IACS). The major interlocutors of scientific modernity in this context were the *bhadraloks*. Furthermore, it is also relevant to appreciate how the various theories of nationalism elaborated by noted scholars help us understand Indian nationalist thought as articulated by *bhadralok* scientists.

Bhadraloks, nationalism, and scientific modernity

The image and practice of modern science, as it developed within colonial India, reflected manifestly conflicting ideological predispositions. On the one hand, there was the Orientalist vision of science as a civilizing mission, espoused by William Jones—a British lawyer, Sanskrit scholar, and the founder of the Asiatic Society—and Lord Macaulay, a British administrator; thus by implication, a vision that had no roots in the Indian culture. On the other hand, there were nationalist *bhadralok* scientists in India in the late nineteenth century who were the mentors of Satyendranath Bose, C.V. Raman, and Meghnad Saha. These mentors included Prafulla Chandra Ray and Jagadish Chandra Bose, for whom cultivating modern science was a route to reviving the glorious tradition of ancient Indian knowledge.

The pioneering effort towards institutionalizing Indian interest in Western science was the founding of the Indian Association for the Cultivation of Science (IACS) in 1876 by Mahendra Lal Sircar, a medical practitioner and well-known social reformer, as previously mentioned. The basic aim of the institute was to encourage Indians in scientific research and to popularize scientific knowledge. In 1895, Sircar remarked:

> We have two kinds of hoarded wealth in this country, one in the shape of hoarded gold and silver, and the other in the shape of unused intelligence. In order to liberate the latter, it is necessary to liberate the former, which in this sublunar world of ours in a magic transformer of energy of all kinds.[61]

Furthermore, in 1897, Satish Chandra Mukherjee, an eminent nationalist and *bhadralok*, launched the *Dawn* magazine, which spread ideas on national education. In 1902, Mukherjee introduced the Dawn Society (as previously mentioned) to promote the concept of national education.[62] Mukherjee's efforts led to the founding of the National Council of Education (NCE), which assisted in keeping science and technology in the curriculum of national education.[63] With the formation of the Indian Science Congress Association in 1914, Indian scientists gained a broad platform for exchanging ideas. A *bhadralok* chemist, Prafulla Chandra Ray, published *A History of Hindu Chemistry* in 1902 as mentioned earlier.

Despite all the political and intellectual ferment in Bengal, a national consciousness was growing that transcended its provincial boundaries. Ray also understood the importance of a nation-wide awakening, saying:

> In these days of awakened national consciousness, the life story of a Bengali chemist smacks rather of narrow provincialism … It will be found, however, that most part of the subject matter is applicable to India as a whole. Even the economic condition of Bengal applies *mutatis mutandis* to almost any province in India.[64]

These developments in nationalizing education, formation of the Indian Science Congress, historical works in the sciences by *bhadraloks*, as well as the presence of a colonial government, created a "nationalist" science. This science was one that very often went beyond the boundaries of the nation, and incorporated ideas from the transnational scene, as will be examined later in this monograph. The *bhadralok* intellectuals eventually forged advances in science, such as the Bose–Einstein statistics, the Saha equation, and the Raman Effect, that were nationalist as well as cosmopolitan in nature. It can be inferred that a nationalist cosmopolitan consciousness was a result of initiatives taken by the *bhadralok* intelligentsia, and also of their interactions with their international colleagues. Science, nationalism, and cosmopolitanism were closely enmeshed, especially with the Indian nationalists who were exposed to local and Western education—the *bhadraloks*.

The existing scholarship and theories of nationalism elaborate how academics have viewed the workings of various forms of imagining the nation. For example,

Ernest Gellner, an eminent scholar of nationalism, has located the "age of nationalism" in the structural transformation of state power, leading to the explanation of a national identity. He argued that industrialization was the primary cause of nationalism.[65] For Benedict Anderson,

> nations were imagined into existence through institutions of print-capitalism in Europe and subsequently appropriated by nationalist elites in Asia and Africa who borrowed the Western "modular" forms of nationalism.[66]

Several Indian nationalist leaders were Western educated and, accordingly, were greatly influenced by Western modular forms of nationalism. Anderson's model has been critiqued by Partha Chatterjee, a scholar who studied anticolonial nationalism, and the subsequent processes of decolonization.[67] Chatterjee pointed out, that if "third world nationalisms were mere emulations of Western models, then even the nationalist imaginations remain colonized forever."[68] This monograph examines further this question of whether nationalist imaginations were colonized forever, as per the thesis of Chatterjee.

This book draws on a wide range of sources and methods, including oral histories, history of scientific ideas in the West and in South Asia, cultural history, intellectual history, postcolonial theory, and archival research using close historical case studies. I see this work as part of a larger effort to make the history of physics and science an integral part of a general South Asian history, and accessible to a wider public. Quantum physics in colonial India, as it was received, understood, and adapted in various ways to local conditions and academic traditions (or a lack of them) outside Europe, is an important area to explore.

India succeeded in developing a strong and original research tradition in modern science while it was a British colony. This success had come a couple of decades prior to India acquiring independence in 1947. Quantum physics held an attraction for, and was subsequently pursued by, a generation of young Indian *bhadralok* scientists who were born and educated in India rather than in Europe. In the chapters that follow, it will be argued why Indian science, through the lens of "*bhadralok* physics," followed such a trajectory; and how quantum physics was received in India.

This monograph describes in detail the methodology for exploring the rise and impact of "*bhadralok* physics" through the case studies of three *bhadralok* physicists: Satyendranath Bose, who is best known for his work with Albert Einstein on the quantum statistics of identical particles; C.V. Raman, who received the Nobel Prize (1930) for his work on the quantum dispersion of light that helped bring about quantum mechanics; and Meghnad Saha, a noted quantum astrophysicist who later helped to establish the institutions of Indian physics, and who worked to persuade Gandhi and Nehru for practical science policies.

This monograph's analyses and conclusions are integrated into the larger field of knowledge. The case studies illustrate and elucidate the origins and conscious emergence of a *bhadralok* outlook among these influential physicists, and its operation as a key component of Indian cultural nationalism. In the final chapter, the monograph integrates my findings into the broader historical understanding and historiographic viewpoints outlined in this opening chapter. These relate, in particular, to an understanding of "*bhadralok* physics" as a worldview and social phenomenon, and its impact on the emergence of Indian cosmopolitan nationalism; which in turn contributed to the making of modern physics in colonial India.

Regarding the sources, there has been a severe dearth of primary sources for the period and characters examined here. This scarcity is part of a larger problem in South Asian history, and presently India is having a difficult time recovering its own history. Because of the Nehruvian developmental model espoused by postcolonial India, and a craving for science, technology, and engineering, studies in the humanities and social sciences experienced a serious setback, and have been reduced to the status of a subordinate.

Moreover, the profession of history in general and the history of science, specifically in India, has almost become a family property. Family members sometimes hold on to primary documents, unwilling to part with them, and often refuse to engage with historians, with few exceptions. For example, I had a hard time locating Raman's spectrograph from the archives. It was only very recently that Rajinder Singh shared with me the photo of Raman's crucial instrument (see Figure 1.1).

An additional obstacle comes from the fact that the field of history of science is not a mainstream area of study or profession in Indian academia. Consequently, scholars in North America working in this area are often seen through a lens of suspicion. Therefore, research in this area is even more challenging and painstaking.

Media sensationalism has also led people to perceive journalists and historians as belonging to the same category. Though I personally have a lot of respect for good journalism, serious scholarship in history and media journalism can never be synonymous. The reason I mention this is because many family members of this

Figure 1.1 Raman's spectrograph.[69]

monograph's protagonists have shown me the door and refused to engage with me, assuming I was a journalist looking for sensationalist, therefore marketable, stories for the media. If this problem can be appropriately addressed, I believe it can open major gateways to furthering research on science in colonial India. I hope that this book will contribute toward this wider goal of refuting stereotypes of Asian scientists, and their respective contributions to the advancement and the making of modern science.

The methodological approach undertaken in this monograph can be characterized as follows. First, this is non-Eurocentric history of science in a colony under the conditions of British Imperialism. Second, this book focuses not on a recent episode (last fifty years), but on how science operated in a period about a century ago. Third, this study uses non-English language sources in exploring the methods and approaches of Indian *bhadralok* scientists. Fourth, it engages with the internal and external context of science, thereby showing that science and culture are deeply entangled. And last but not the least, it connects the separate fields of South Asian history and the history of science, thereby starting to bridge an important historiographic lacuna in the existing body of scholarship. The narrative and analysis in the forthcoming chapters will follow the three *bhadralok* scientists, their approaches and methods of doing physics and the culture in which they were bred.

Notes

1 On a preliminary level, the Higgs Boson discovery was made on July 4, 2012 and a more recent confirmation on March 14, 2013.
2 Somaditya Banerjee. "Transnational Quantum: Quantum Physics in India Through the Lens of Satyendranath Bose." *Physics in Perspective* 18, 2 (2016) 157–181.
3 Sumit Sarkar. *Writing Social History* (Delhi, 1998), 169 as quoted in Tithi Bhattacharya. "In the Name of Culture" *South Asia Research* 21, 2 (2001) 161–187.
4 Ibid.
5 John McGuire. *Making of a Colonial Mind: A Quantitative Study of the Bhadralok in Calcutta, 1857–1885.* (Canberra: Australian National University Press, 1983) 18–31, 42–83.
6 N. Jayaram. "The Making of a Colonial Mind: A Quantitative Study of Bhadralok in Calcutta, 1857–1885" [book review]. *Contributions to Indian Sociology* 19, 206–207.
7 Amit Kumar Gupta. *Crises and Creativities: Middle-Class Bhadralok in Bengal, c. 1939–52.* (Hyderabad: Orient Blackswan, 2009).
8 Ibid.
9 Ibid. 8.
10 Swati Chattopadhyay. *Representing Calcutta: Modernity, Nationalism, and the Colonial Uncanny.* (London and New York: Routledge, 2006) 138–139.
11 I thank Robert Anderson for this clarification.
12 Dhruv Raina and S. Irfan Habib. "The Moral Legitimation of Modern Science: Bhadralok Reflections on Theories of Evolution." *Social Studies of Science* 26, 1 (1996) 9–42.
13 Ibid.
14 Historians based at Jawaharlal Nehru University (New Delhi) and others who espouse Jawaharlal Nehru's vision.
15 See Chapter 5 on Meghnad Saha.

16 Fritz Ringer. *The Decline of the German Mandarins: The German Academic Community, 1890–1933*. (Hanover: Wesleyan University Press, 1990).

17 See Note 18.

18 Russell McCormmach. "On Academic Scientists in Wilhelmian Germany." *Daedalus* 103, 3 (1974) 157–171; and Fritz Ringer. *The Decline of the German Academic Community, 1890–1933*. (Hanover: Wesleyan University Press).

19 McCormmach, 158.

20 Ibid. 160.

21 Robert Brain. "Bürgerliche Intelligenz." *Stud. Hist. Phil. Sci.* 26, 4 (1995) 617–635.

22 Peter Galison and Gerald Holton. *Einstein for the Twenty First Century: His Legacy in Science, Art and Modern Culture*. (Princeton: Princeton University Press, 2008) 3.

23 Robert Brain. "Bürgerliche Intelligenz." 619.

24 Paul Forman. "Weimar Culture, Causality, and Quantum Theory, 1918–1927: Adaptation by German Physicists and Mathematicians to a Hostile Intellectual Environment." *Historical Studies in Physical Sciences* 3 (1971) 1–115; "Scientific Internationalism and the Weimar Physicists: The Ideology and its Manipulation in Germany after WWI." *Isis* 64 (1973) 151–180.

25 Mario Biagioli. *Galileo Courtier: The Practice of Science in the Culture of Absolutism* (Chicago: University of Chicago Press, 1904). Steven Shapin and Simon Schaffer. *Leviathan and the Air-Pump*. (Princeton: Princeton University Press, 1989). Geoffrey V. Sutton. *Science for a Polite Society: Gender, Culture, and the Demonstration of Enlightenment*. (Boulder: Westview Press, 1997).

26 Peter Galison. *Image and Logic: A Material Culture of Microphysics*. (Chicago: University of Chicago Press, 1997); Peter Galison. *Einstein's Clocks, Poincare's Maps: Empires of Time*. (New York: W.W. Norton, 2004); Andrew Pickering. *Constructing Quarks: A Sociological History of Particle Physics*. (Chicago: University of Chicago Press, 1994).

27 "Contact zones" is a term which was coined in the intellectual history landscape by Mary Louise Pratt in 1991.

28 Abha Sur. *Dispersed Radiance: Caster, Gender and Modern Science in India*. (New Delhi: Navayana, 2011) 25.

29 Mara Beller. *Quantum Dialogue: The Making of a Revolution*. (Chicago: University of Chicago Press, 2001).

30 As Camilleri argues, "what we now refer to as the Copenhagen Interpretation (CI) had its origins in discussions between Niels Bohr and Werner Heisenberg in the latter part of 1926 and early 1927." Bohr's idea of complementarity, his radical ideas supporting causality, wave theory, and his measurement postulate are usually regarded as the central ideas of the CI. K. Camilleri. "Constructing the myth if the Copenhagen Interpretation". *Perspectives on Science*, 17 (2009) 26–57.

31 Richard Staley. *Einstein's Generation: The Origins of the Relativity Revolution*. (Chicago: University of Chicago Press, 2008); "On the Co-creation of Classical and Modern Physics." *Isis* 96, 4, (2005) 530–558.

32 Russell McCormmach. "H.A. Lorentz and the Electromagnetic View of Nature." *Isis* 61, 4 (1970) 459–497.

33 For a similar argument see Theodore Arabatzis' article on "The Electron's Hesitant Passage to Modernity 1913–1925." In M. Epple and F. Muller (eds.). *Science as Cultural Practice*. (Berlin: Akademie-Verlag, 2017).

34 Frederick Cooper. *Colonialism in Question: Theory, Knowledge, History*. (Berkeley: University of California Press, 2005) 113–152

35 Ibid.

36 Partha Chatterjee. *Nationalist Thought and the Colonial World: A Derivative Discourse*. (Minneapolis: University of Minnesota Press, 1993) 3–13; *The Nation and Its Fragments: Colonial and Postcolonial Histories*. (Princeton: Princeton University Press, 1993).

37 Partha Chatterjee. *Our Modernity* (Senegal). (Rotterdam and Dakar: CODESRIA-SEPHIS, 1997) 3–20.
38 In the context of my study, South Asia means India.
39 Kapil Raj. *Relocating Modern Science: Circulation and the Construction of Knowledge in South Asia and Europe, 1650–1900.* (New York: Palgrave Macmillan, 2007) 229.
40 Non-elite voices from people not included in the narratives of a nation or non-dominant social groups who are marginalized.
41 Vivek Chibber. *Postcolonial Theory and the Specter of Capital.* (London: Verso, 2013) 33.
42 Antonio Gramsci. *Selections from the Prison Notebooks*, ed. and trans. Quintin Hoare and Geoffrey Nowell Smith. (London: Lawrence & Wishart, 1971) 324. Ranajit Guha. *A Subaltern Studies Reader, 1986–1995.* (Minneapolis: University of Minnesota Press, 1982) 1–8.
43 Homi Bhabha. *The Location of Culture.* (London and New York: Routledge, 1994). Not to be confused with the senior Indian nuclear physicist Homi Bhabha.
44 John Lourdusamy. *Science and National Consciousness in Bengal, 1870–1930.* (New Delhi: Orient Longman, 2004) 5–33.
45 The IACS formed by Sircar was a response to the system of reserving membership of the Asiatic Society (founded in 1784) to only the British for the first few decades of its working.
46 Lourdusamy, *Science and National Consciousness*, 141.
47 Pratik Chakrabarti. "Review of John Lourdusamy, Science and National Consciousness in Bengal." *Medical History* 50, 3 (2006) 403–404.
48 Ibid.
49 Subrata Dasgupta. *Jagadish Chandra Bose and the Indian Response to Western Science.* (Oxford: Oxford University Press, 2000).
50 Ibid.
51 Ibid.
52 Ibid.
53 In a personal conversation with the author in March 2010.
54 Robert Anderson. *Nucleus and Nation: Scientists, International Networks and Power in India.* (Chicago: University of Chicago Press, 2010) 7.
55 Ibid. 17.
56 Ibid. 538–539.
57 Ibid. 538–539.
58 Deepak Kumar. "*The Making of the Indian Atomic Bomb: Science, Secrecy, and the Postcolonial State.* Itty Abraham." *Isis* 92, 1 (March 2001) 213–214.
59 Prakash Kumar. *Indigo Plantations and Science in Colonial India.* (Cambridge and New York: Cambridge University Press, 2012) 9.
60 Gyan Prakash. *Another Reason, Science and the Imagination of Modern India.* (Princeton: Princeton University Press, 1999) 200–203.
61 Uma Dasgupta, ed. *Science and Modern India: An Institutional History c. 1784–1947.* (New Delhi: Centre for Studies in Civilizations, 2011) 69–117.
62 Pratik Chakrabarty. *Western Science in Modern India: Metropolitan Methods, Colonial Practices.* (New Delhi: Permanent Black, 2004) 12.
63 Uma Dasgupta, ed. *Science and Modern India: An Institutional History c. 1784–1947.* (New Delhi: Centre for Studies in Civilizations, 2011) 849–870.
64 Prafulla Chandra Ray. *A History of Hindu Chemistry.* (2 vols). (Calcutta: Chuckervertty, Chatterjee & Co., 1902–08); *Life and Experiences of a Bengali Chemist.* (Calcutta: Chuckervertty, Chatterjee & Co., 1935).
65 Ernest Gellner. *Nations and Nationalism.* (Ithaca: Cornell University Press, 1993).
66 Benedict Anderson. *Imagined Communities: Reflections on the Origin and Spread of Nationalism.* (London and New York: Verso, 1983).

67 Benedict Anderson, *Imagined Communities* 17–49. For a critique of Benedict Anderson, see Partha Chatterjee, *The Nation and Its Fragments: Colonial and Postcolonial Histories.* (Princeton: Princeton University Press, 1993) 3–13.

68 Partha Chatterjee. *The Nation and Its Fragments: Colonial and Postcolonial Histories.* (Princeton: Princeton University Press, 1993) 3–13.

69 I thank Rajinder Singh for this photograph.

2   *Bhadralok* culture and the making of Satyendranath Bose

In 1924, a thirty-year-old unknown Indian physicist from Calcutta by the name of Satyendranath Bose (1894–1974) wrote a short letter to the then famous, forty-five-year-old German physicist, Albert Einstein (1879–1955), in which Bose requested assistance with the publication of his paper entitled "Planck's Law and Light Quantum Hypothesis." Although Einstein had little idea who the author was, he read the paper, translated it into German, and forwarded it to the German journal *Zeitschrift für Physik* for publication. Regarding Bose's paper, Einstein said: "In my opinion Bose's derivation of the Planck's formula constitutes an important advance. The method used here also yields the quantum theory of the ideal gas as I shall discuss elsewhere in more detail."[1]

Einstein was quite pleased with Bose's novel derivation of the Planck's Law, and read this paper at the Physico-Mathematical Colloquium in the Berlin Academy of Sciences.[2] He immediately sent a letter of praise to Bose, calling his work a beautiful step forward.[3] After sending his letter, Einstein then extended Bose's approach from light quanta to material gas. Their collaboration by correspondence formed the basis for the foundation of a novel concept in physics that became known as Bose–Einstein statistics, or simply Bose statistics. The correspondence between Bose and Einstein is a special moment in the history of science, because Bose's paper had already been rejected from the prestigious *Philosophical Magazine*. It is possible that because Einstein was Jewish—a peripheral identity in Weimar Germany—Bose resonated with him to the extent that, in spite of Bose's communication originating from the peripheries of science in a British colony, it was sufficient for the further development of physics that led to the Bose–Einstein statistics. It is also notable that Einstein had attained global celebrity status by the early 1920s, especially after the Eddington eclipse expedition. Hence, his inspirational image was something to be emulated and respected by a scientist from a colony like Bose's.

Bose's original paper, along with Einstein's subsequent one, influenced the work of Erwin Schrödinger, and contributed to the creation of the new quantum mechanics in 1925. In 1926, following Bose's method, Enrico Fermi, and later Paul Dirac, derived a new distribution formula for an assembly of particles, obeying Pauli's exclusion principle. This formulation was later known as the Fermi–Dirac statistics. Additionally, particles that obeyed Bose–Einstein statistics were called "Bosons," a name coined by Dirac in the third edition of his *Principles of Quantum Mechanics* published in 1947.[4]

Bose's 1924 letter to Einstein transformed Bose's career from that of a relatively unknown scientist in the colonial world to an active participant in the ongoing revolution in modern physics. A well-known biographer of Einstein, Abraham Pais, calls Bose's 1924 paper "the fourth and last of the revolutionary papers of the old quantum theory (the other three being, respectively Planck's, Einstein's and Bohr's)," adding "I believe there had been no such successful shot in the dark since Planck introduced the quantum in 1900."[5] Hence, even though Bose's contribution became famous through Bosons, Bose–Einstein statistics, and the Higgs Boson, his role has remained largely unknown not only to physicists, but also to people like Pais, who wrote and published on Einstein and the history of the quantum revolution. While Bose is mentioned in passing in narratives of quantum physics by Mara Beller and Emilio Segre, there is no mention of him in recent works of history of physics, like Suman Seth's *Crafting the Quantum*, or Jed Buchwald and Andrew Warwick's *Histories of the Electron*.[6]

Over the last two decades, cultural historians of science have conducted a wealth of research emphasizing the local embedding of knowledge production. In a variety of case studies, these scholars have demonstrated how supposedly universal scientific knowledge is generated in local contexts, and how it retains the specific cultural fingerprints of its origin.[7] They have also analyzed the ability of local knowledge to travel and spread beyond its place of origin. A number of recent transnational approaches to cultural studies of science have focused on the processes of translation, diffusion, and transformation, the crossing of cultural boundaries, and the global circulation of the locally embedded scientific knowledge.[8]

This chapter belongs to this recent trend and takes up the case of one of the "hardest" sciences—quantum physics—which originated in Germany during the first three decades of the twentieth century. Historians have analyzed the social and cultural contexts of late Imperial and Weimar Germany, and the ways in which they contributed to the development of quantum physics.[9] However, they have not sufficiently analyzed examples of transnational knowledge that flows horizontally, offering better indices of knowledge interchange, and instead have focused on a model that presumes a vertical relationship between a center and a periphery.[10]

Furthermore, the interconnectedness between Bose and Einstein depicts Indian science as a complex form of cultural hybridization between the local and the global, including the broad notion of a *Visvajaneenata* cosmopolitanism. By cosmopolitanism I mean a non-hierarchical mode of coexistence of the local and the global in landscapes with a power differential.[11] Cosmopolitanism implies an interconnection between the local and the universal, with an intellectual ethos espousing a vision of a culturally embedded global scientific consciousness. Moreover, by local *Visvajaneenata*[12] cosmopolitanism I mean a synergistic cross-pollination between the localities of scientific knowledge, which are born in a specific cultural context, and myriad strands of transnational thought. As will be shown later, for example, Bose–Einstein statistics and bosons are examples of this type of local *visvajaneen* cosmopolitanism.

Postcolonial theorists widely use the concept of hybridity, which usually refers to the creation of transcultural forms within the space produced by colonization. It usually identifies the crossbreeding of two species by grafting in order to develop a third "hybrid" species. In the South Asian context, hybridity has been used by postcolonial theorist Homi Bhabha.[13] The problem with this kind of analysis is that when applied to science in South Asia, it appears to be rather simplistic. In contrast to "'hybridity,'" the notion of local cosmopolitanism is broader and more applicable. I argue that scientists like Bose espoused a unique brand of local cosmopolitanism that often combined traditional Indian culture influenced by British traits, with features that were neither Indian nor British, showing the transnational spectrum of the notion of local cosmopolitanism.[14]

Born to a lower-middle-class family in 1894 in Calcutta, then the capital of British India, Bose was the family's only male child. His father, Surendranath Bose, worked for the colonial government. Bose "had an aptitude for mathematical thinking and showed interest in several branches of science" (Figure 2.1).[15]

Bose happened to be his parents' eldest child. Consequently, he received a great deal of attention from his father because of the patriarchal social structure in India in the nineteenth and early twentieth century, where having a male child was like a boon for the family. In the face of hard work and his family's meager wealth, Bose's father took the time to see that his son received a quality education.

One may infer from Surendranath's life trajectory that he disliked the colonial government, considering that he eventually left his job with the railways (which

Figure 2.1 Bose as a student circa 1910–11.[16]

were owned and controlled by the government), and started a modest chemical and pharmaceutical company with his associate Satish Chandra Brahma in 1901. What is more, some of Bose's early childhood memories reflect witnessing the emerging nationalist movement, as communicated to us by his close friend Melvyn Brown:

> One morning circa 1900 father and son were walking towards Surendranath's workplace when the air suddenly exploded with voices. Satyen stopped, though his father tugged at his hand to move on. The voices rose louder as a group of young Bengalis took the turning at the corner, and came face to face with the guardians of the law. Let us watch them, father! No – this is not the time for it. Why are they shouting? You're too small to understand, son. Is it something to do with the English, father? Yes, father replied.[17]

Bose's family belonged to the Kayastha caste, (a subcaste below the Brahmins found in Bengal), who did not have traditional access to education and academia.[18] The field of education had traditionally belonged to the monopoly of the Brahmins, who were known for their scholarship, especially in Sanskrit. Yet by the end of the nineteenth century, Brahmins were losing their age-old grip on the sphere of education, which had gradually opened up to members of other castes. Historians have characterized this process as a consequence of the Bengal Renaissance that started in the early nineteenth century, with the appearance of a large number of newspapers, periodicals, the growth of numerous societies and associations, and several reform movements through which people in Bengal found ways to publicly discuss their problems, including the impact of British rule on the Indian subcontinent.

Although debates continue in current historiography regarding the appropriateness of the concept of the Bengal Renaissance, many scholars share the view of Subrata Dasgupta that intellectual development and distinctive collective cognitive identity in nineteenth-century Bengal can be characterized as a certain "Renaissance phenomenon."[19] Other scholars, for example David Kopf, suggested that the traditional meaning of the term "Renaissance" as a rebirth of culture is not strictly applicable, but can be used as an intellectual tool if understood as the process of change and adaptation of cultural values and attitudes, which was a product of nineteenth-century cosmopolitanism.

This reawakening led to the rise of the *bhudraloks,* and is somewhat similar to the German "mandarins," what Russell McCormmach calls *Kulturträger* ("culture-bearer"), or the Prussian *Bildungsbürgertum* (educated members of the German bourgeoisie) in late nineteenth-century Wilhelmian Germany.[20] This movement served at once to foster national culture, to divert support from current political authorities, and to promote its adherents into the upper social and scientific strata. As Gerhard Sonnert argues, "*Bildungsbürger* were people who had received a Gymnasium and a university education and who worked predominantly in professions that required training, such as physicists, lawyers, clergy, teachers, and professors, as well as other higher officials in government service (*Beamte*)."[21]

As such, *bhadraloks* believed that their work was the *raison d'être* of the nation. While this similarity with German intellectuals is important, it is also interesting to note that the Indian scientists were all early-career scholars in their early twenties, unlike Wilhelmian academic scientists who were usually older than the average scientist. While this rise of the German intelligentsia happened after the Franco–Prussian war (1871), the rise of the *bhadraloks* happened for the most part after the Sepoy Mutiny (1857).

Using the functioning of the College of Fort William (1800) as an example, Kopf emphasizes the role of British Orientalism and Indian intellectual culture. The newly emerging "public sphere" opened possibilities for all castes to contribute to learned culture (*Kultur*) and academic matters, and allowed those like Bose's father, Surendranath, a non-Brahmin Kayastha, to aspire for education for himself, and even more strongly for his only son. Through these aspirations, Surendranath's family became part of the new social group in colonial India often called the *bhadraloks*.[22]

Bhadraloks, the Indian intelligentsia

Among the major consequences of the Bengal Renaissance was the creation of the new Indian intelligentsia, the *bhadraloks*. The term itself is a Bengali word, but was applicable across India as a new identity starting in the 1830s. *Bhadralok* as a label was especially popular among the growing Indian middle class (*madhyabitta*), many of whom worked in district towns and were employed by local governments along with the colonial administrations.[23] One could find people of different backgrounds among this newly defined group that was comprised of landowners, industrialists, professionals, bureaucrats, teachers, poets, novelists, and freelance writers. Although partially created and educated to fulfill the needs of the colonial governance, this growing "middle-class," while drawing income from the administration, formed the social base of the reform movements, and participated energetically in the development of the new print culture, contributing to the rise of nationalist mentality.[24]

Bhadraloks should not be considered a class or a homogenous group, but could be subdivided into three categories. The first included those privileged ones who "were in high offices" of the British. The second can be characterized as the middle-class proper, who were not "rich but comfortable." The third included the relatively poor "lower middle-class, who were nevertheless *bhadra*," (i.e., similar to the middle class in education, fashion, and manners).[25]

Whether landed rentier class or petty bourgeoisie, *bhadraloks* shared a common attachment to the value of education, which could be conceptualized in different ways. At the same time, *bhadraloks* also often acted as incipient spokesmen for the nation. They initially emerged as part of the cultural and literary reawakening in early nineteenth-century Calcutta,[26] the capital of British India. But the movement soon spread beyond the confines of Bengal, particularly to Benaras in the north, and to Pune in the west.

The Indian intelligentsia that participated in this social and intellectual activity developed a growing awareness, and a pride for the Indian past, especially in the high traditions of Indian philosophy. But *bhadraloks* took a distinctively different form from the version of that past, which was maintained by the traditional educated caste of the Brahmin pundits with their mastery of Sanskrit language and classical literature.[27]

Sanskrit, the classical language of ancient India, "was seen by the British as a secret language invented by the Brahmins to be a mysterious repository of their religion and philosophy."[28] There had always been a considerable curiosity about the religion of the *Gentoos*[29] amongst the Europeans, and some of them had made efforts to learn Sanskrit. "Whatever knowledge the British had about the scholasticism and religious thoughts of the Hindus came from discussions with Brahmins and other high caste Indians. Brahmin pundits were professors and some even came to be conceived of as lawyers."[30]

Bhadralok intellectuals occupied an intermediate stratum between the traditional intellectual elite, the Brahmanical scholasticism of the Sanskrit pundits, and a majority of indigenous population that was devoid of literacy altogether. They also had an ambivalent attitude toward European culture in the widest sense of the term, embracing it with a mixture of anxiety and aspiration as something to be partially emulated and partially rejected. As a result, *bhadraloks* developed a unique brand of cosmopolitanism that mixed Indian traditional culture with some British influences, and sometimes also with features that were neither Indian nor British. We may characterize this ideological synthesis with the help of a deliberately self-contradictory term, "cosmopolitan nationalism," or *Visvajaneen Jatiyatabaad*, in Bengali.

One of the milestones in the emergence of *bhadraloks* was the founding of the College of Fort William by Lord Wellesley in 1800.[31] The primary aim of the college was to give a similitude of European education and training to Indian natives, and to produce clerks who would help the British in administration. The Indian staff that was recruited for the college included a number of distinguished scholars, such as Mrityunjay Vidyalankar from Midnapore, who made major contributions to the prose literature of Bengal.[32] Some of the staff made a few noteworthy intellectual contributions and influenced their European counterparts.

It is important to clarify a couple of interpretative points with regard to the *bhadralok* category. This group, by virtue of its access to European-style education, transcended caste and class barriers. Several prominent *bhadraloks* hailed from a lower caste and a lower class.[33] These included Radhakanta Deb, Meghnad Saha, Bijoli Behari Sarkar, Girindra Sekhar Bose, Rasiklal Dutta, Ashutosh Dey, Sarasilal Sircar, and Pramatha Nath Bose. When we address the question of the emergence of the first modern-style scientists in late colonial India, it is important to understand that they were identified and self-identified as part of this wider group of *bhadraloks*. "High status" was not a necessary part of their background and family, but rather an aspiration. One had to study and work hard to be considered a *bhadralok*; for example, a janitor ("*Abhadra*"[34] or Not-*Bhadra*) could also be identified as a *bhadralok* if he read Tagore's poems.[35]

The defining characteristic of the *bhadraloks*, in my view, was that, even as they were major harbingers of modernity in colonial Indian society, they tended to reject or to deviate from both the Orientalist attitudes of the British colonizers, and the traditionalist attitudes of the Indian Brahmin elite, especially in the following crucial aspect. Instead of strongly separating modern science from traditional knowledge, the *bhadraloks* were inclined to combine one with the other.

The making of a modern Indian scientist

Bose's upbringing was conservative, typical of a middle-class family of the early twentieth century. However, his father placed special emphasis on education. Bose started elementary school at age five. At the outset, he went to Normal School, which used to be close to his father's rented house in Jorabagan in Calcutta. When his family moved to their own house at Goabagan, Bose entered the neighboring New Indian School. From the above narrative of events it can be inferred that Bose's father was interested in getting Bose trained in a more competitive atmosphere. So, his father sent him to Hindu school, where Bose studied English, Bengali, History, Geography, Mathematics, and Sanskrit. Interestingly enough, the prescribed textbook in mathematics used to be Gauri Sankar Dey's *Arithmetic and Algebra*. At Hindu school, Sarat Chandra Shastri, the Bengali teacher, infused in Bose's mind a passion for Bengali language and literature.

In 1905, when Bose was eleven years old, the Partition of Bengal, enacted on July 19 of that year by Lord Curzon, the Governor General of India, sparked nationalist sentiment in India. Partially in response to the growing uprisings in Bengal, which had alarmed the colonial authorities, the British administration decided on a plan to thwart the movement by dividing Bengal into two separate regions—Western Hindu and an Eastern Muslim province—in the interest of a "diminution of the power of Bengali political agitation."[36] The event marked the beginning of a new phase in the history of Indian nationalism. The struggle against the partition of Bengal led to the beginnings of the *Swadeshi* (indigenous) movement, which encouraged domestic production and the boycotting of British goods. Political figures like Surendranath Banerjee and Bal Gangadhar Tilak were key members of the Indian National Congress (INC), who helped reshape people's conceptions of the probable trajectory of India's independence.

Slogans like *Bande Mataram* (Hail Motherland), coined by Bankim Chandra Chattopadhyay (Chatterjee), conjuring up the image of Goddess Durga, became the national cry for freedom and were chanted within schools, colleges, and other nationalist circles.[37] Growing up in a conservative lower-middle-class family, the teenage Bose was specifically ordered by his father to stay away from revolutionary activities. Although Bose conceded to his father's wish, he sympathized with the revolutionaries and often thought about them.[38] His family's influence, which included a strict father, a loving mother, and younger sisters, molded him, at least in his youth, to restrain the expressions of his political sentiments.

Bose's friend, Meghnad Saha, chose a different path and got expelled from Dacca Collegiate School at age twelve, when he, along with some of his classmates, staged a boycott of the visit of the Governor during the time of the Partition of Bengal in 1905.[39] Saha had always been actively involved with the Bengal revolutionaries before Indian independence (1947), and embarked on a political career in the latter part of his life (the 1950s). By contrast, Bose's nature showed a manifest shyness, a sense of modesty, and a seemingly docile character. Though Bose's inclinations were similar to that of a nationalist, his childhood days and the economic conditions he experienced precluded him from engaging in active revolutionary activities, in spite of the lasting impression left on his mind by the Partition of Bengal. Bose recalled that impression along with his childhood memories many years later in a Convocation address given at the Calcutta University in 1973. When he spoke of the formative period of his life, his earliest memory seemed to go back to the year 1905 and the protests against the Partition of Bengal.[40]

In 1905, Bengal was partitioned by Lord Curzon. This was a very significant moment in the history of Indian nationalism, as it aroused patriotic feelings far and wide. During this time, Bose became involved in several student protests. Nationalist chants like *Bande Mataram* were used persuasively in Bengal to let the British know how deeply the residents of both East and West Bengal were hurt. Bose was especially influenced by this manifest geographical discontinuity, whereby the Muslim majoritarian, East Bengal was separated from a Hindu majoritarian, West Bengal, on seemingly "administrative" grounds because of Curzon.

The patriotic teachings of social reformers, like Ram Mohan Roy, Vivekananda, and noted litterateur Bankim Chandra, took on a concrete shape and became sources of lasting inspiration for people. Nationalists across India took up Bengal's cause and were uniformly appalled at the British arrogance, and what appeared to be blatant tactics of "divide and rule".[41] The *Swadeshi* movement kindled a spirit of patriotism in Bose, along with many others.[42] The song *Bande Mataram* (Hail Mother), composed by Bankim Chandra Chatterjee, became the informal anthem of the nationalist movement after 1905, and a battle cry of Indian nationalists. The opening words of the song are as follows:

Mother, I bow to thee!
Reach with thy hurrying streams,
Bright with thy orchard gleams,
Cool with thy winds of delight,
Dark fields waving, Mother of might,
Mother free ...
Who hath said thou art weak in thy lands,
When the swords flash out in twice seventy million hands ...
To thee I call, Mother and Lord! ...
Thou art wisdom, thou art law,
Thou our heart, our soul, our breath,
Thou the love divine, the awe
In our hearts that conquers death

Every image made divine
In our temples is but thine.[43]

Bose entered the Hindu School in 1907. Girijapati Bhattacharya, a childhood friend of Bose, remarks:

> Our friendship began in 1908, when we were both at the Hindu School, Calcutta, though he was a year ahead of me. Even at school, Satyendranath was marked for his extraordinary intellect. Our mathematics teacher predicted that Satyendranath would one day be a great mathematician like Laplace or Cauchy ... In fact, he once gave him 110 out of 100 in mathematics because not only did he get all the sums right, he had done some of them in more than one way![44]

Bose brought home a report card where the grade of 110 had been given to him out of 100 for a mathematics test. When his father decided to point out this blunder to the mathematics teacher, Bose's teacher said, "No, Sir, there has been no mistake ... he [Satyendranath] has done all the sums correctly ... including the alternatives, all within the appointed time."[45]As he considered his father to be his mentor, his role model, and his *guru*, Bose wanted to learn science just like his *guru*. His training, however, revealed some interesting features. The prescribed textbooks were *Arithmetic and Algebra* by Gaurisankar Dey, *Grammar* by Rowe and Webb, *Jungle Stories* by Kipling, and a few other novels by Vidyasagar.[46]

Bose's education was eclectic, because he got a chance to read both Indian and Western authors, who might have given him some idea about differences in writing styles. Despite his weak eyesight, Bose was a thoughtful reader. His favorite poets were Tennyson and Rabindranath Tagore, and he was well versed in Kalidasa's *Meghadootam* in Sanskrit. At Hindu school, he showed signs of his unfolding linguistic intellect through his natural liking for languages, especially Sanskrit, French, and later, German. However, Bose wanted to take the entrance exam for Intermediate Science[47] in 1908. But, as he was sick, he was unable to do so, and kept studying at the same school. In this brief period, in 1908, he took the time to familiarize himself with advanced mathematics, and a few Indian classics in Sanskrit before joining Presidency College in 1909.

Taking into account the cultural and intellectual milieu in which Bose grew up, one can think of two probable and interrelated reasons why Bose decided to study science. The first reason was the strong influence of his father, as well as the norms and idioms that Bose had internalized. The other reason was the rising nationalist movement, coupled with the impact of the Bengal Renaissance. Because of his scholarly and contemplative bent of mind, and also because he was reluctant to join the British administration, especially in the aftermath of the Partition of Bengal, Bose opted to pursue an academic career in teaching, instead of service to the Government, as his father Surendranath had done in his early life. Speaking in the Calcutta University in 1973, Bose recalled the days of his youth:

At that time (1910s) the school-days were not over—then came the high tide of patriotism. In teens, we wandered in the streets by singing the songs of Rakhibandhan. We wanted to feel we all are brothers; we all are children of the India, irrespective of castes and religions. We have to remove the distresses of our poor India, have to bear the striving of bondage of the foreign rules – we have to revive a great nation of old tradition from the cruel exploitation and rule of foreigner. We have to inspire the people with old ideas to modern thoughts – we have to drive away illiteracy. Friends, well fifty years ago when we were young we had one great idea that we had to prove to the world that *Indian science* is not less than anybody else and therefore we were anxious to show our own intelligence, our originality … they were great things on those days … at that time, those of us who opted for science as the first preference were able to do something for the nation.[48]

The tradition of the Rakhibandhan[49] mentioned above was started by the nationalist leader, Surendranath Banerjee. It involved a custom of tying a colorful wristband (*Rakhi*) to one another as a demonstrative protest against partition. The wristband had to be tied to one's left wrist as a symbol of Indian unity. The day chosen for the ceremony of Rakhibandhan was the day on which Partition was proclaimed in 1905. As people met one another on that day, each person had to tie a *Rakhi* around someone's wrist. The image of an indigenous yellow string had a powerful appeal for the imagination of the youth that Bose belonged to.

Bose was not alone in reversing the traditional career priorities by choosing an academic trajectory over colonial administrative service.[50] He was just one of a generation of students from Bengal for whom nationalism and pursuit of an academic calling went hand in hand and became closely linked. The likeminded students from the well-known 1909 graduating class of Presidency College, Calcutta, were Meghnad Saha, Jnan Chandra Ghosh, Nikhil Ranjan Sen, Jnanendranath Mukherjee, and Pulin Bihari Sarkar. The science faculty of Presidency College had a distinguished staff, and included Prafulla Chandra Ray in chemistry, Jagadish Chandra Bose in physics, and D.N. Mallik, and C.E. Cullis in mathematics.[51]

As Mehra remarks,

> Bose and his classmates shared the (double) excitement of acquiring scientific knowledge and the patriotic fervor derived from the *Swadeshi* movement. They wanted to put scientific knowledge to use through technology for the benefit of the masses. Bose took the B.Sc. final examination in 1913 and received the M.Sc. degree in Mixed Mathematics (similar to applied mathematics) from Calcutta University in 1915. He ranked first in both examinations, second place going to Meghnad Saha.[52]

Sailen Ghosh came first in Physics in M.Sc. in the same year that Bose came first in Mixed Mathematics—indicating a novel mix of physics and mathematics in the Indian academic profession of the time.

Figure 2.2 A group of *bhadralok* intellectuals in the 1910s. Seated (left to right): Meghnad
Saha, Jagadish Chandra Bose, Jnanchandra Ghosh. Standing (left to right):
Snehamoy Dutt, Satyen Bose, Debendramohan Bose, Nikhil Ranjan Sen,
Jatindra Nath Mukherjee, N. Chandra Nag.[55]

As Irene Gilbert argues, one can trace the origin of the Indian academic profession by following the Wood's Despatch (1854), which declared the educational
policies of the government of the East India Company. A possible answer to the
often-asked question as to why academics lacked the potential for a regular liveli-
hood in nineteenth-century India lies in the consequence of "the dominance of the
Indian Civil Service (ICS) over the Indian Educational Service (IES),"[53] which
(ICS) included the science and engineering profession. Members of the ICS did
not have much respect for academic pursuits, nor did they "promote research or
accord much respect to members of the academic profession" (Figure 2.2).[54]

First appointment and training

In 1916, Bose received his first academic appointment as lecturer in the Applied
Mathematics department at Calcutta University. At the time, Ganesh Prasad was
the Ghosh professor of Applied Mathematics at Calcutta University. Prasad had
received the first D.Sc. to be awarded by Allahabad University in 1898, and later
earned his Mathematical Tripos from Cambridge. He then served as a professor
of mathematics at the *Kayastha Pathsala* at Allahabad, before joining Calcutta
University in 1916.[56] Prasad taught in Queen's College at Banaras before coming

to Calcutta. He also worked with Felix Klein and Hilbert at Göttingen.[57] Bose, however, did not get along with Ganesh Prasad:

> The students flocked to him (Prasad) for training in research. They were the best science students of Calcutta though several of them had not secured high marks in Ganesh's paper in their M.Sc. But the fault lay with the teachers at Presidency College- at least that is what Ganesh Prasad thought. The young students had to stomach adverse comments about their former (Indian) teachers, too scared to answer back. After my M.Sc. I too presented myself before Ganesh Prasad who was also my examiner though I had not fared as badly as the others. Dr. Prasad was kind to me at first but I was notorious for plain speaking. I found it notorious to bear his tirade against my teachers. I had dared to counter his adverse criticisms. This infuriated him. He said – you may have done well in the examination but that does not mean you are cut out for research. Disappointed I came away. I decided to work on my own.[58]

Prasad hailed from the North of India, and, possibly for the reason of regional cultural differences, did not get along with Saha and Bose. As the following chapters will explain, regionalism and the conflict between different regional identities became a recurring phenomenon in the development of Indian science, sometimes hindering its institutionalization, and the work of scholars. A manifestation of regionalism would be an overemphasis on theoretical explorations in science (too many theoretical physicists fighting for greater personal recognition) in different regional centers, as opposed to group collaboration in experimental work. Some of these trends continued after independence, and still exist in twenty-first-century India. In Bose's case, due to his conflict with Prasad, he decided to transfer from mathematics to the physics department of the same university.

At the time, the physics department of Calcutta University had a dearth of teaching faculty, and Bose had to take the lead in teaching and organizing the department. His transition from a graduate student to the unexpected appointment as professor in physics created an opportunity for him to teach a graduate seminar, and get acquainted with the most recent state of research in theoretical physics. The education he received, and the textbooks that were available in India at the time, provided inadequate bits of information regarding the ongoing radical scientific developments in European physics, involving relativity, quantum, and atomic theories. Scientific journals arrived irregularly due to the Great War. However, the Presidency College library collections did have the *Philosophical Magazine*. This is important, as Bohr's pathbreaking papers on the atomic model, published in the *Philosophical Magazine*, were available to the patrons of the Presidency College library. Bose also started learning French and German languages in order to be able to read the European scientific literature that he could access.

Following his master's degree examination in 1915, Bose continued his studies in physics and applied mathematics at the newly established University College, Calcutta.[59] At this time, P.J. Brühl, an Austrian, taught engineering physics at Bengal Engineering College. Brühl's book and journal collections consisted of a vast array, which included recently published works in the "old quantum theory" and relativity. Bose recalled:

> Since Saha and I had learned some German, we were glad to borrow these things from Bruhl. He possessed a good collection of advanced texts and journals on physics in German. He had Planck's *Theorie der Waermestrahlung*, Laue's *Das Relativitaetsprinzip*, as well as papers on quantum theory and relativity.[60]

With the help of Debendramohan Bose (D.M. Bose)[61], a physicist who had just returned from Germany, Bose also gained access to Max Planck's lectures on thermodynamics, originally published in 1897. Furthermore, he also received a copy of Gibbs's treatise, *Elementary Principles in Statistical Mechanics*, published in 1902, from where Bose learned more about the concepts of phase space in the context of statistical mechanics. Ashutosh Mukherjee also had a wide array of scientific texts which were important for Bose's further learning of physics concepts.[62]

Through these incidental pieces of available literature, Bose gradually became acquainted with the exciting and counterintuitive developments in recent physics, but his very isolation from the physics community in Europe also made him oblivious to some of the skeptical views still lingering in Europe at the time. Luckily, he remained unaware of the critique of Einstein's light quantum, which, prior to the discovery of Compton Effect (1923), was typically rejected by the existing authorities in physics. This lack of information ultimately proved advantageous for Bose's own work in the field.

The reception of relativity in India and the making of Bose

Soon after the end of World War I, the Indian public started receiving exciting news about the ongoing revolution in physics. In 1919, newspapers in Britain triumphantly announced the experimental verification of Albert Einstein's general theory of relativity by the British astronomer Arthur Eddington.[63] The Calcutta newspaper, *The Statesman*, sent a reporter to the astronomical observatory at the Science College in Calcutta University, to obtain a lay explanation of a cabled confirmation of Einstein's prediction of deflection of starlight in the gravitational field of the sun.[64] Responding to this request, Bose and Saha translated several of Einstein's papers on special and general relativity, which were published the following year by the University of Calcutta Press as a book titled *The Principle of Relativity*.[65] This book happened to be the first translation of Einstein's seminal papers into English.

In India, as elsewhere, the excitement about relativity, space, and time spread far beyond the professional community of scientists, and affected the general public, philosophers, and literati. In his novel, *Shesher kabita*,[66] the Nobel laureate poet Rabindranath Tagore referred to Einstein's relativity with a philosophical interpretation, finding that Einstein's notion of relative simultaneity resonated in the Indian cultural context in that: "Time should not mean the same to everybody. Conventional clock gives one time relative to space, but personal clock which controls the Universe, gives another. This is what Einstein thinks."[67] During this time, the Indian science community was still quite small, and the Indian physics and astronomy communities were even smaller. Their reaction to and reception of relativity was noticeably different from that of the British. First, the separate discipline of theoretical or mathematical physics had not established itself in India, and most Indian physicists combined experiments with mathematical calculations. This relative lack of preexisting tradition in theoretical physics turned into an advantage for them when it came to the reception of relativity, since Indian physicists, unlike their British counterparts, did not need to overcome a strong attachment to the traditional concepts of classical physics, such as ether. If anything, beliefs regarding ether had been perceived in India with skepticism even before relativity; for example, Swami Vivekananda, the Indian philosopher and social reformer, wrote in 1895:

> As far as it goes, the theory that this ether consists of particles, electric or otherwise, is also very valuable. But on all suppositions, there must be space between two particles of ether, however small; and what fills this inter-ethereal space? If particles still finer [sic], we require still more fine ethereal particles to fill up the vacuum between every two of them, and so on. Thus the theory of ether, or material particles in space, though accounting for the phenomena in space, cannot account for space itself. And thus we are forced to find that the ether which comprehends the molecules explains the molecular phenomena, but itself cannot explain space because we cannot think of ether as in space. And, therefore, if there is anything which will explain this space, it must be something that comprehends in its infinite being the infinite space itself. And what is there that can comprehend even the infinite space but the Infinite Mind?"[68]

Pointing to the philosophical deficiencies of the theory of ether in connection with the concept of infinite space, Vivekananda's remark reveals ambiguities concerning how ether was understood in India—especially in particulate or continuous terms—even though Indian scientists were certainly aware of it through English education and textbooks.

Prior to 1919, Einstein and his relativity concept found little notice in India; and there was a skeptical response from those who noticed, which elicited minimal popular expression.[69] Speaking in 1922, at the Presidential Address of the Indian Science Congress Association, (Madras), British scientist C.S. Middlemiss remarked:

> I must say that, though still intensely inquisitive in the matter of this high and elusive doctrine (of relativity), I must reluctantly conclude that no help is to be derived from such of the popular attempts at explanation as I have so far seen. Ordinary scientists, the unfortunate plain man and the practical person

have no chance here I'm afraid. It would seem that there is no royal road to understand Relativity. It must be approached by the same laborious track that has been responsible for its inception and development, namely, by the way of Higher Mathematics.[70]

Middlemiss' above remark shows how British scientists reacted to Einstein's theory; and how, when examined from a British industrial perspective, relativity did not make much practical sense, since the former espoused "Higher Mathematics."

Einstein's scientific methodology was quite philosophically diffuse. This pragmatist view emerges from Einstein's childhood thought experiment of racing with a ray of light, and seeing a spatially periodical electromagnetic field at rest, which neither experience nor Maxwell's equations would allow. Other examples include his subsequent magnet-conductor thought experiment, where Einstein jettisoned ether; as well as the moving-train experiment, by virtue of which Einstein concluded that observers moving with respect to one another disagree over whether or not two events at different places are simultaneous (relativity of simultaneity).[71]

Though Einstein's general theory of relativity was not empirically driven, experience did confirm it.[72] In fact Einstein's attitude towards empirical data could be summarized from the following remark at a lecture in Berlin:

> The theorist's method involves … general postulates or "principles" from which he can deduce conclusions …The scientist has to extract these general principles from nature by perceiving in comprehensive complexes of empirical facts certain general features which permit of precise formulation.[73]

Indian scientists responded to Einstein's theory in a way different from the British since, although their British-style education was gradually getting them acquainted with the British point of view, they were not thoroughly embedded in British industrial culture. Conceptual differences also trickled down because of the language of science. The language barrier played a role in these differences as a handful of Indian scientists knew German. They relied mostly on British textbooks for their training and education; and in Britain, too, recognition for Einstein's work was only granted very slowly. During the 1910s and 1920s, students graduating from Cambridge or Oxford were not required to know about relativity theory. Additionally, the curriculum at the University of London, a leading center of physics, was not required to offer any courses on the subject.[74]

In their studies regarding the reception of the theory of relativity, Andrew Warwick and Richard Staley show that there was only a gradual development of the understanding, that relativity constituted a radical break from the existing theories of ether. Early reactions to Einstein's new theory were mixed and slow in coming. In 1905, Einstein was an obscure patent clerk in Zurich, whose career up to that point showed little indications that he was about to turn the world of physics on its head. Initially, many mathematical physicists regarded Einstein's contribution as just another paper, phrased in obscure language, on the electrodynamics of moving bodies.

The science magazine *Nature*, for example, mentioned Einstein's views on relativity in the same breath as those of Cambridge-trained physicist Joseph Larmor, and ether theory's foremost champion, Oliver Lodge. German-trained physicists, who were more sympathetic to the tradition of research in which Einstein had been trained, were more receptive to the possibilities that his theory of relativity unlocked. One of the first to respond positively to Einstein's theory was Max Planck, who presented a seminar on Einstein's theory in 1905 in Berlin. Einstein himself published a series of papers over the next few years, expanding and refining his theory. One of these papers contained his first proofs of the famous equation linking mass and energy, $E = Mc^2$, stating that the energy of a body is equal to its mass multiplied by the square of the speed of light.[75]

As a result of dabbling with magnet-conductor-type experiments as Einstein did, Lucasian Professor Joseph Larmor at Cambridge believed in the existence of an absolute frame of reference in the form of electromagnetic ether. As Andrew Warwick argues, "Larmor also continued to believe that the equations of electromagnetism were not themselves fundamental but were ultimately to be derived from the dynamical properties of the underlying ethereal medium by the application of the Principle of Least Action."[76]

It gradually became clear to Cambridge Maxwellians that Einstein had abandoned the concept of ether entirely, but they tended to treat this as a metaphysical rather than physical stance. In the Cambridge tradition of the early twentieth century, a conceptualization of physics without ether was virtually impossible; and Cambridge physicists found it hard to accept or take seriously the claim that no form of ether existed.[77] The persistence of ether in Cambridge illustrates the power of a conservative training regime in resisting the novel ideas in theoretical physics coming from Germany.[78] It is possible that the level of mathematics training of Indian scientists —albeit a weak one—made it possible for them to be aware of relativity before 1919.

Given how slowly relativity was acknowledged in England, it is not surprising that it took even longer for it to arrive in the British colony of India. One notable exception was Calcutta University, which was the leading center of physics in early-twentieth-century India. A nearly complete stoppage of circulation of non-German journals during the Great War complicated matters even further.[79] Without exaggeration, one can thus say that relativity arrived in India only in 1919, with a dramatic announcement of the astronomical confirmation of general relativity by Eddington. It was general relativity that most attracted the public

Even the collection of translations by Bose and Saha, and the introduction to the volume by Indian scientist Prasanta Chandra Mahalanobis, paid relatively little attention to the special theory of relativity, but gave a much greater coverage to insights and explanations about general relativity. The reception of relativity theory in India brought with it a new disciplinary way of executing science. Bose and Saha were not only the first to translate Einstein's original papers into English, they also came to be viewed as the *first theoretical physicists in India*, whose work and research strategies followed the model of this new branch of science, which had already been firmly established in Germany and in Central Europe.

It is interesting to note that Bose and Saha's translation was more precise than the somewhat later British translation of Einstein's papers. An archival letter sent in 1993 by the famous historian and philosopher of science, Max Jammer, to Bose's student Partha Ghose at Kolkata explains this point very clearly. Max Jammer was replying to a fax inquiry by Ghosh about Bose and Saha's translation of Einstein's paper "On the Electrodynamics of Moving Bodies."

Though German was not the native language of either Bose or Saha, they overcame the language barrier and produced a translation that was error free, while the British translators (Jeffery and Perrett) made a small language mistake that resulted in a serious distortion of the physical meaning of the text. Jammer's letter[80] reconfigures some of the stereotypes about colonial scientists' inability to produce knowledge of the same level as metropolitan scientists.[81] This example of transnational knowledge flowing horizontally offers a better index of knowledge interchange, and helps to challenge a model that presumes a vertical relationship between a center and a periphery.

Hence, Bose's worldview can be framed as a locally rooted cosmopolitanism, or *Visvajaneenata*; he was bound locally by his training in India, characterized by its distance from the centers of Maxwellian wave theories in England; yet he could also exploit the universality of physics to reach out to a person of Albert Einstein's esteem. This cosmopolitanism of scientific culture, which allows a scientist from a colony to enter into a dialogue with key scientific figures in the metropole, is an important feature of my study, and helps bridge the local and the global through narratives of science.

Many existing biographies of Bose belong to the hagiographic genre that unfortunately remains a problem in the history of science.[82] However much one may admire Bose's scientific contributions, this chapter's overarching aim is not so much about the celebration of Bose but rather the understanding of the historical events, which happened in a very specific temporal and geographical setting. In order to achieve this narrative, I examine Bose's life as a scientist attracted to his field by the growing nationalist movement, by the lure of modernity, and as a participant in the web of debates within Western science. I have described Bose's social role as a *bhadralok* scientist, belonging to a distinctly new and evolving group in late colonial India.

Notes

1 SNBCS Archives Doc 15 (Satyendra Nath Bose Centre for Basic Science, Kolkata).
2 Mahadev Dutta, *Satyendra Nath Bose: Mathematician, Scientist & Humanist*. (Calcutta: Calcutta Mathematical Society, 1995) 6.
3 Jagdish Mehra and Helmut Rechenberg. *The Historical Development of Quantum Theory. Vol. 1: The Quantum Theory of Planck, Einstein, Bohr and Sommerfeld. Its Foundation and the Rise of Its Difficulties (1900–1925)*. (New York: Springer Verlag, 1982) 565.
4 Paul Dirac. *Principles of Quantum Mechanics* 3rd ed. (London: Oxford University Press, 1947) 210.
5 Abraham Pais. *Subtle is the Lord: The Science and Life of Albert Einstein*. (London: Oxford University Press, 2005) 425–428.

6 Suman Seth. *Crafting the Quantum: Arnold Sommerfeld and the Practice of Theory, 1890–1926.* (Cambridge, MA: MIT Press, 2010); Jed Buchwald and Andrew Warwick, eds. *Histories of the Electron: The Birth of Microphysics.* (MIT Press, 2004).

7 Mario Biagoli and Galileo Courtier. *The Practice of Science in the Culture of Absolutism.* (Chicago: University of Chicago Press, 1994); Steven Shapin and Simon Schaffer. *Leviathan and the Air Pump: Hobbs, Boyle and the Experimental Life.* (Princeton: Princeton University Press, 1985).

8 Peter Galison. *How Experiments End.* (Chicago: University of Chicago Press, 1987); Andrew Pickering. *The Mangle of Practice: Time, Agency and Science.* (Chicago: University of Chicago Press, 1995); Richard Staley. *Einstein's Generation: The Origins of the Relativity Revolution.* (Chicago: University of Chicago Press, 2009); Somaditya Banerjee. "Transnational Quantum: Quantum Physics in India through the lens of Satyendranath Bose." *Physics in Perspective* 18, 2 (August 2016) 157–181.

9 Paul Forman. "Scientific Internationalism and the Weimar Physicists: The Ideology and its Manipulation in Germany after WW1." *Isis* 64 (1973) 151–180.

10 Deepanwita Dasgupta. "Stars, Peripheral Scientists and Equations: The Case of M. N. Saha." *Physics in Perspective* 17, 2 (June 2015) 83–106.

11 Pheng Cheah. "Cosmopolitanism." *Theory, Culture, & Society* 23, 2–3 (May 2006) 486–496.

12 A Bengali word which means "cosmopolitanism."

13 Homi Bhabha. *The Location of Culture.* (London: Routledge, 1994).

14 Claude Markovits. "How British Was British India." *Jahrbuchfur Europaische Uberseegeschichte* 10 (2010) 67–91.

15 Jagdish Mehra. "Satyendranath Bose. 1 January 1894–4 February 1974." *Biographical Memoirs of Fellows of the Royal Society* 21 (November 1975) 118.

16 SNBCS Archive Doc No. 0117.

17 Melvyn Brown. *Satyendranath Bose.* (Annapurna Publishing House, 1974).

18 Chitrarekha Gupta. *The Kayasthas: A Study in the Formation and Early History of a Caste.* (Calcutta: CK.P. Bagchi, 1996). Caste is a system of social stratification in India. Indian society is characterized by a unique system of stratification called caste which divides society in four different *varnas* or stratas, namely Brahmin (the priestly and teaching caste), Kshatriya (the warrior caste), Vaishya (the trading caste), and lastly the *shudras* (the servile caste). Because of the unique stratification system, Indian society has often been considered a closed society, which does not permit upward (vertical) social mobility for the lower castes. See Pitirim Aleksandrovich Sorokin. *Social Mobility.* (New York and London: Harper and Brothers, 1927) 139. As it is today, the Indian caste system exhibits the following characteristics. Birth determines the caste of an individual. Merits and/or achievements do not enable one to elevate oneself from a lower caste to a higher one. On the contrary, any failure to conform to caste norms may lead to the degradation of a person from a higher caste to a lower one. The caste system is endogamous, that is, the members of caste marry within their own caste group. Caste exclusiveness is not confined to marriage alone but embraces almost all areas of social life. Furthermore, caste distinctions are displayed in surnames, so that the caste of a person can be immediately known from the surname. For example, "Saha" which is a surname, serves as the lowest caste identification mark (*shudra*). Pitirim Aleksandrovich Sorokin. *Social Mobility.* (New York and London: Harper and Brothers, 1927).

19 Subrata Dasgupta. *The Bengal Renaissance: Identity and Creativity from Rammohan Roy to Rabindranath Tagore.* (Permanent Black, 2006) 2–65; David Kopf. *British Orientalism and the Bengal Renaissance.* (Berkeley: University of California Press, 1969) 11–30.

20 Russell McCormmach. "On Academic Scientists in Wilhelmian Germany." *Daedalus* 103 (Summer 1974) 157–171; Fritz Ringer. *The Decline of the German Mandarins: The German Academic Community, 1890–1933.* (Harvard, 1969).

21 Gerhard Sonnert. *Einstein and Culture*. (Humanity Books, 2005) 52. Also see Sam Schweber. *Nuclear Forces: The Making of the Physicist Hans Bethe*. (Harvard University Press, 2012).

22 David Kopf. *British Orientalism* 280–289; S. Samanta. *The Bengal Renaissance: A Critique*. Paper presented at the European Conference of Modern South Asian Studies, Manchester, UK (July, 2008) 2. Samanta gives a nice commentary here of Kopf's argument about the Bengal Renaissance.

23 B.B. Misra. *The Indian Middle Classes: Their Growth in Modern Times*. (New York: Oxford University Press, 1961); Pradip. Sinha. *Nineteenth Century Bengal: Aspects of Social History*. (Calcutta: Firma K.L. Mukhopadhyay, 1965).

24 Gyan Prakash. *Another Reason: Science and the Imagination of Modern India*. (Princeton: Princeton University Press, 1999).

25 Tithi Bhattacharya. *Sentinels of Culture: Class, Education, and the Colonial Intellectual in Bengal (1848–85)*. (New York: Oxford University Press, 2005) 38.

26 David Kopf. *British Orientalism and the Bengal Renaissance: The Dynamics of Indian Modernization 1773–1835*. (University of California Press, 1969).

27 For example, Pandit Ganesh Datta Shastri. And also, Persian Maulvis like Maulvi Muhammad Nasir al- Din Haidar. See for example: Richard M. Eaton. *Rise of Islam and the Bengal Frontier 1204–1760*. (University of California Press, 1996) 213.

28 Bernard Cohn. *Colonialism and its Forms of Knowledge: The British in India*. (Princeton: Princeton University Press, 1996) 25.

29 An eighteenth-century name given by the British referring to a Hindu. See for example Nathaniel Brassey Halhed. *A Grammar of the Bengal Language*. (Bengal: Hoogly, 1778) x– xiii.

30 Sheldon Pollock. *The Language of Gods in the World of Men: Sanskrit, Culture and Power in Premodern India*. (University of California Press, 2006) 75–88.

31 Nitish Sengupta. *The History of the Bengali Speaking People*. (UBS, 2002).

32 Cohn. *Colonialism and Its Forms of Knowledge* 50.

33 By lower caste I mean non-Brahmins; by lower class I mean economically poor.

34 *Abhadra* means "Non-gentlemanly" and *bhadra* means "gentlemanly".

35 This reminds me of the work of Pierre Bourdieu. *Distinction: A Social Critique of the Judgement of Taste*. (Routledge, 1986.) In essence, how do aesthetic choices create class-based social groups? The question then arises as to how is this *bhadralok* identity spatialized? Though it is difficult to answer, this new identity was seen more in an urban setup, but rural locale could also produce *bhadraloks*. An example is that of Meghnad Saha who was born in the village of Seoratali in Dacca and grew up to be one of most famous *bhadralok* scientists India has ever produced.

36 As stated by Lord Apthill, a chief administrator in Calcutta, in 1903. See Bidyut Chakrabarty. *The Partition of Bengal and Assam, 1932–1947*. (Routledge, 2004) 87.

37 Sumit Sarkar. *Modern India 1885–1947*. (Macmillan India, 2007) 83.

38 Oral interview with Enakshi Chatterjee at Kolkata, June 2012.

39 Robert Anderson. *Nucleus and Nation: Scientists, International Networks, and Power in India*. (Chicago and London: The University of Chicago Press, 2010).

40 Partha Ghose. "Bose Statistics: A Historical Perspective." In Majumdar et al. (eds.). *S. N. Bose: The Man and His Work*. (Calcutta: S N Bose National Centre For Basic Sciences, 1994) 11.

41 Thomas Metcalf. *A Concise History of Modern India*. 3rd ed. (Cambridge: Cambridge University Press), 157.

42 Mehra. "Satyendranath Bose" 117–154.

43 Metcalf. *A Concise History of Modern India* 156–157.

44 Partha Ghose. "Bose Statistics: A Historical Perspective." In Majumdar et al. (eds.). *S N Bose: The Man and His Work* Pt 2, 9–11.

45 Brown. *Satyendranath Bose* 29.

46 Santimay Chatterjee and Enakshi Chatterjee. *Satyendra Nath Bose*. (Calcutta: National Book Trust, 2005) 10.

47 Hayden J. A. Bellenoit. *Missionary Education and Empire in Late Colonial India 1860–1920*. (Pickering & Chatto, 2007). This book will be using Bose, Satyen Bose, SNB as variants for referring to Satyendranath Bose.

48 S.N. Bose Archives (Document Number 49, 50). *Calcutta Mathematical Society*, Kolkata. (Accessed June 15, 2010).

49 Indian ritual (during colonial India) of binding a decorated string on one's wrist as a symbol of brotherhood.

50 Enakshi Chatterjee. "The Business of Freeing India." *The Statesman*, Calcutta, January 9, 1994.

51 Santimay Chatterjee. *Satyendranath Bose* 12.

52 Mehra, "Satyendranath Bose" 119.

53 Irene Gilbert. "The Indian Academic Profession: The Origins of a Tradition of Subordination." *Minerva* 10 (1972) 384.

54 Ibid., 384.

55 SNBCS Archives, Kolkata Doc. 0067 (accessed July 2012).

56 *Hindustan Review* and *Kayastha Samachar* VIII (November 1903) 466.

57 B.N. Prasad. "Obituary: Prof Ganesh Prasad, His Life and Work." *Science and Culture* I (August 1935) 142–145.

58 Santimay Chatterjee. *Satyendranath Bose* 23.

59 Santimay Chatterjee. "Satyendranath Bose" 21–30; M.N. Saha, "Obituary: Dr. Brühl." *Science and Culture* I (October 1935).

60 Mehra. "Satyendranath Bose" 120.

61 Nephew of Jagadish Chandra Bose.

62 Mehra. "Satyendranath Bose" 120–122.

63 Katy Price in *Loving Faster than Light* argues that "the one key feature of the new space and time that stood out was that almost nobody could understand or explain it." Katy Price. *Loving Faster than Light: Romance and Readers in Einstein's Universe*. (Chicago: University of Chicago Press, 2012).

64 Robert Anderson. *Nucleus and Nation: Scientists, International Networks and Power in India*. (Chicago: University of Chicago Press, 2010).

65 A. Einstein and H. Minkowski. *The Principle of Relativity*, trans. M.N. Saha and S.N. Bose. (Calcutta: University of Calcutta, 1920).

66 Sisir K. Majumdar. "Rabindranath's Thoughts on Science." *Frontier* 44 (2011) 53–54.

67 Ibid.

68 Swami Vivekananda. "The Ether." *New York Medical Times* (February 1895) 58. A discussion of how ether was conceptualized in ancient India is beyond the scope of this monograph. But this debate surrounding ether's status has often been traced in British and German understandings in this period.

69 Meghnad Saha and Satyendranath Bose. *The Principle of Relativity.* (Calcutta: Calcutta University Press, 1920) 1–33, 89–154.

70 C.S. Middlemiss. "On Relativity." In *The Shaping of Indian Science: Indian Science Congress Association Presidential Addresses, Vol 1: 1915–1947*. (Hyderabad: Universities Press, 2003) 101–107.

71 Banesh Hoffman. *Albert Einstein: Creator and Rebel*. (New York: Penguin Books, 1972) 9–11.

72 Jean Eisenstaedt as quoted in Thomas Glick. "Cultural Issues in the Reception of Relativity." In *The Comparative Reception of Relativity*. (Dodrecht: D. Reidel Publishing Company, 1987), 381–400; L. Fang. *China and Albert Einstein: The Reception of the Physicist and his Theory in China, 1917–1979* [book review]. *The China Journal* (2006) 55, 211–212; Danian Hu. *China and Albert Einstein*. (Harvard University Press, 2005); Danian Hu. "The Reception of Relativity in China." *Isis*, 98,

(2007) 539–557. These pieces give a good commentary on how relativity was received in China and the Japan connection.

73 Inaugural lecture, Berlin, July 2, 1914. See *Collected Papers of Albert Einstein. Vol.6, The Berlin Years: Writings 1914–1917*, ed. A.J. Kox, Martin J. Klein, and Robert Schulmann (New Jersey: Princeton University Press, 1996).

74 Andrew Warwick. *Masters of Theory: Cambridge and the Rise of Mathematical Physics* (Chicago: University of Chicago Press, 2003) 360.

75 Staley (2008) 4–12 as reviewed by Stanley (2009) 470–471. Richard Staley. *Einstein's Generation: The Origins of the Relativity Revolution.* (Chicago: University of Chicago Press, 2008) 4–12. Matthew Stanley. "Einstein's Generation: The Origin of the Relativity Revolution" [book review]. *The British Journal for the History of Science* 42 (2009) 470–471. Staley argues how Einstein's relativity theory was dependent on several factors. Stanley's positive review of Staley corroborates the argument of Staley about how the histories of relativity written by actors during the period under study was contingent on various events and seemingly non-linear.

76 A. Warwick. "On the Role of the Fitzgerald-Lorentz Contraction Hypothesis in the Development of Joseph Larmor's Electronic Theory of Matter. *Archive for the History of Exact Sciences* 43 (1991) 33.

77 Though Warwick claims that British electromagnetic theory was not primarily about ether and originated in continuum mechanics. So, there are some qualifications to the statement. But three outstanding features of German electrodynamics in the mid-1890s were the belief in the existence of electromagnetic ether, lack of a physical picture of the electrical current, and fascination for electrodynamics of moving bodies.

78 Olivier Darrigol. "The Electrodynamic Origins of Relativity Theory." *Historical Studies in the Physical and Biological Sciences* 26, 2 (1996) 241–312; Russell McCormmach, "Lorentz and the Electromagnetic View of Nature" *Isis* 61(1970) 459–461.

79 Robert Anderson notes as per personal communication with N.G. Barrier that "the British did not ban German scientific literature, at least in the Indian theatre." See Anderson. "Nucleus and Nation" 597, footnote 18.

80 Ibid.

81 SNBCS archives Kolkata, Document SL. Nos. 33, 34 (accessed July 2012).

82 Melvyn Brown. "Satyendranath Bose"; Jagadish Mehra. "Satyendranath Bose"; Santimay and Enakshi Chatterjee. *Satyendranath Bose*; and Mahadev Dutta. *Satyendranath Bose*.

3   Satyendranath Bose and the concept of light quantum

Using the work of Indian physicist, Satyendranath Bose as a lens, this chapter will explore how fundamentally new concepts of German quantum physics transformed and established roots in different cultural and political circumstances, namely the conditions of colonial India. Additionally, it explores how a physicist from colonial India shaped German physics by establishing Bose–Einstein statistics.

As this chapter will display, Bose's derivation of the Planck's Law vindicated a view Einstein had championed for roughly nineteen years. Nor was it just a "shot in the dark" that triggered Bose's insight, as Abraham Pais has suggested. Bose was very much aware that his result was a logical development of Einstein's work, an insight that had eluded Einstein himself for nearly two decades. Nevertheless, Bose was working in an isolated fashion in a remote colony—standing apart from Maxwellian physics—where he was dependent on texts and journals circulated from Europe by émigrés such as P.J. Brühl. As a result, Bose not only knew how to appeal to common thought within the physics community, but also was isolated enough to be free from the temptation of rejecting Einstein's outlook in order to appease scientific orthodoxy. Furthermore, in order to appreciate Bose's approach, the idea of local *Visvajaneenata* cosmopolitanism helps flesh out how a colonial scientist, working within a power differential, generated new knowledge, and engaged with a metropolitan scientist like Albert Einstein.[1]

Local *Visvajaneenata* cosmopolitanism of Bose's science

Even prior to his involvement with the translation of Einstein's papers, Bose showed a special interest in German physics. This interest was obviously not an insignificant matter during the Great War, when Germany and Austria were the British Empire's primary enemies, and at a time when German-language science was boycotted by Britain and the country's allies. Bose's disobedience and connections to German science and continental works (because of the relative availability of such resources), which reflected his sense of alienation from the Empire and from British colonial rule, was part of a more general pattern that one can characterize as local *visvajaneen* cosmopolitanism in the emergence of Indian science.

Other chapters will show how Indian scientists pursued international collaborations beyond the confines of the British Empire, such as between Bose and Einstein, Raman and Arnold Sommerfeld, and Saha and Walther Nernst, among others. An obvious consequence of the cosmopolitan dimension was that it allowed Indian scientists, living within the confines of a colonial regime, to develop research agendas that were independent of the lines of research pursued in the center of the British Empire. For Bose, the idea of the quantum provided a great intellectual escape from the hegemony of scientific colonialism.

The new quantum physics, which originated primarily in Germany, had an even stronger appeal for Indian scientists than relativity theory, because it allowed them to pursue experimental work as well as theoretical work.[2] Thus, it is no accident that the emerging Indian physics particularly excelled in the novel discipline of quantum physics. On the surface, the quantum did not appear directly relevant in the colonial Indian context; nor was quantum physics imported to India from Britain. Both British physics and the British-style education available in India at the time worked according to the paradigm of the Maxwellian continuum universe.

Quantum physics, by contrast, represented the microscopic world and embodied a very discontinuous worldview, seen as very radical and counterintuitive, especially from the perspective of classical physics.[3] We can understand the appeal, the meanings, and the importance of the quantum for Indian science, when we examine Bose's contribution to quantum physics, especially through its local cosmopolitan and non-colonialist aspects. Despite his origins in a remote Asian colony, Bose managed to master the cutting-edge research in this new field, and contributed to its further development through his unique quantum statistics.

In 1924, when Bose was teaching at Dacca University, he had applied for a two-year research sabbatical to visit Europe. Though there were bureaucratic obstacles to overcome in order to get this leave, Einstein's handwritten postcard (July 2, 1924) in response to Bose's first letter (June 4, 1924), in which Einstein commented favorably on Bose's re-derivation of Planck's Law, helped to expedite the sabbatical process. Hartog (see Figure 3.1), the vice-chancellor of Dacca University, helped facilitate the travel logistics and gave Bose a generous research allowance. Bose remarked:

> As soon as [the Senate Council] showed it to Hartog [the Vice-Chancellor], it solved all problems. As a student Hartog had spent some time at the University of Paris and he understood something of what such an experience could do for a young man. That little thing [the postcard from Einstein] gave me a sort of passport to the study leave. They gave me leave for two years and rather generous terms. Then I also got a visa from the German consulate just by showing them Einstein's card. They did not require me to pay the fee for the visa![5]

Sailing from Bombay, Bose arrived in Paris in October of 1924. While visiting Europe, he significantly avoided going to Britain, but rather chose to spend his

Figure 3.1 Philip Hartog.[4]

time in France and Germany, where he established contacts with physicists working on quantum topics. An important recollection by his close friend, Jacqueline Eisenmann, who he met in Paris, sheds light on Bose's goals and motivations. Eisenmann wrote the following brief account in a letter (and also an interview) to Bose's student, Purnima Bose in the summer of 1980:

> I was then a young girl who had just finished her "licence de sciences physiques" and who had just begun to work in Professor Cotton's lab ... Sylvain Levi the great Indianist and Sanskritist was a friend of my father (Dr. Leon Zadoc-Kahn who was in 1943 assassinated by the Germans with my mother). Learning from my father that I intended to work in Physics, Levi said he would make me know "un jeune physician genial".
>
> I was very impatient to meet this genius. When he came to my lab, accompanied by another Indian named Tendulkar, he did not tell me so as to tease me, who was the physicist. Bose was so unassuming that I didn't find out immediately who was who! From that day, I saw him very often. He always went to Paul Langevin's lectures. Langevin gave many lectures. Louis de Broglie came later, Langevin told Madame Curie about him. Bose worked in Madame Curie's lab and in Maurice de Broglie's lab for some time. He went

very much to the museum, loved nature, particularly the Alps, went to see and live in the countryside.

He talked much about Bengali … writing science in Bengali – to teach the students in Bengali. He impressed me very much by his great love for his country. He never went to England until India was free. In 1953, he went to England and lived with Dirac.[6]

After Paris, Bose then traveled to Germany in October of 1924, perhaps because he found that he could communicate more efficiently with German scientists.[7] Writing to Eisenmann from Berlin in 1926, Bose remarked (see Appendix A.1):

I am in my new rooms since the first day I arrived here; it is very nice and comfortable and I am really in love with the balcony … my friends live very near me, about 5 minutes walk from here, in the very buildings of the laboratory and I go there almost every day …

Everybody (every physicist) seems to be quite excited in Berlin, about the way things have been going on with physics, first on the 28th last, Heisenberg spoke in the colloquium about his theory, then in the last colloquium, there was a long lecture on the recent hypothesis of the spinning electron (perhaps you have heard about it). Everybody is quite bewildered and there is going to be very soon a discussion of Schrödinger's papers. Einstein seems quite excited about it. The other day coming from the colloquium, we suddenly found him jumping in the same compartment where we were, and forthwith he began to talk excitedly about the things we have just heard. He has to admit that it seems a tremendous thing considering the lot of things which these new theories correlate and explain, but he is very much troubled by the unreasonableness of it all. We are all silent but he talked almost all the time, unconscious of the interest and wonder that he is exciting in the minds of other passengers.[8]

This letter shows Bose's excitement and enthusiasm while he was in Germany along with his participatory nature as a scientist. As the new quantum mechanics was unfolding in Germany, astonishment also grew because of the counterintuitive nature of the new formalisms that were developing in physics. On Bose's nationalistic frame of mind, Eisenmann remarked:

On 25 July 1973, I asked Professor Bose why he had gone to Paris in the first place. He answered, "I was informed that my friend Abani Mukherjee (a terrorist nationalist leader who was absconding) was in trouble. I had taken some money for him from the country. After meeting Abani, I thought that I will stay in Paris for a while. I had many friends there. They asked me to stay on, I got the idea of doing some experiments so that I would teach these to students in our country. I worked for a while in Maurice's lab. He had already read my paper. He told me that his brother did the kind of work that I did."

When I asked him about his encounters with Einstein in Berlin, he said, "You know that Einstein was included towards "red" – so that chauvinistic German students used to create trouble for him. This made him abandon class lecture. We used to go to his house. He had no research student either. He used to tell us what he thought and sometimes gave lectures too."[9]

Subsequently, regarding Bose's personality, Eisenmann remarked:

It was a great joy to know Bose at all. He was so wonderful, so gifted, knew so much about Hebrew literature and religion. He had an extraordinary heart! He had nearly feminine reaction! He had no ambition for himself, too modest and humble a young man. Referring to a letter by Bose, she remarked, "The letter written in 1951 … was sent a few days after we met in Paris after being entirely without news since 1929."[10]

Eisenmann continued to remark on what Bose did scientifically after returning to India from Europe by saying:

He told me in Paris, after the war (in 1951), as I asked him why he had not published more work, that his surroundings were not favorable. He added, he had spent a great deal of time in preparing experimental research work for his pupils in Dhaka.

Moreover, he said another time that he threw away most of his works that he (Einstein) judged not good enough.

Among his total 24 [sic] published papers, Professor Bose had published 17 papers after coming back from Europe, between 1936 and 1955. The context and content of these papers have not yet been analyzed. The topics range from mathematics, theoretical and experimental physics to biochemistry. Why some of them were not of fundamental importance towards the progress of physics? They were the outcome of resolving obstacles in handling problems by his students and friends, and also on problems which are of practical use for our country.

Sometimes he used to spend days after days in chemical laboratories. Under his guidance students were able to prepare some useful medicines and also to make significant contributions in the synthesis of important chemical compounds.

During the fifties, Bose followed Einstein in his research on "Unified Field Theory". Several other very eminent scientists such as Herman Weyl, Kaluza, and Schrödinger were also involved in the field for many years. Bose was drawn into this movement of handling very difficult mathematical problems and contributed five papers during 1953– 55 which were published in French scientific journals.

Self-taught Meghnad and Satyendranath were teaching untrained students in newly established Departments of Science. one of them knew the art of getting an aim and building up towards it, and the other expressed ideas like flashes of lightening from clouds spread all over in random clusters. Before their partnership could gather sufficient momentum, one was transferred to Allahabad and the other to Dhaka. Ours is a large country and each region has its own specific demands. Only a handful of men had to cope with the growing demands. They were always under pressure concerning everything. In the West, the scene was different.[11]

The problem of the language of science in international, colonial, and postcolonial settings occupied Bose's thoughts throughout his life. Bose felt that science had to play a major role in the service of the Indian nation, but how this could be achieved was a question that often troubled him. He felt that the reason why modern science had not been successful in making progress in India was the hindrance of the medium of instruction. As such, a foreign language was not very appropriate for articulating Indian modernity. Bose had once participated in a global seminar on the relationship of science and culture in Japan, where Japanese (and not English) was the medium of instruction in the entire event. This came as a shock to him. He remarked:

The Japanese use plenty of loan words, but they are not apologetic about it … It is often said as an excuse that lack of Indian synonyms may act as a handicap (in translating from English to Bengali). I am not a purist. I welcome the idea of using English technical terms … We have a lot of such words of foreign origin which have now been absorbed in the regional languages. Everybody understands what is meant by railway, telegram, centimetre, wheel, thermometer, bacteria, fungus, etc. Table and chairs are part of our life now. There is no need to lengthen the list.[12]

In this instance, Bose's personality manifested a certain brand of local *Visvajaneenata* cosmopolitanism. In this context, the term local *Visvajaneenata* that I have coined here describes the way in which Bose went beyond the boundaries of a nation in response to his scientific temperament, and the colonial situation in which India found itself. Nonetheless, *Jatiyatabaad* (nationalism) and *Visvajaneenata* (cosmopolitanism) were not antagonistic to one another in terms of Bose's outlook. His pursuit of science was primarily motivated by a desire to make his countrymen familiar with modern science and its concomitants.

This patriotic sentiment characterized his outlook, and prompted him to use light quantum and German thinking as an escape from his colonial situation. Yet Bose incorporated it in the Indian scientific tradition by his original work in quantum statistics. Therefore, his nationalistic aspirations transcended the boundaries of the nation. He was prepared to appropriate newer concepts (such as quantum discontinuity) for the progress of science in general, and essentially set up an ordered system of ideas through his statistics, while doing away with the apparent contradictions between *Jatiyata* (nationalism) and local *Visvajaneenata* (cosmopolitanism). Hence, the deliberate coinage of the apparently self-contradictory term *Visvajaneen Jatiyatabaad* (cosmopolitan nationalism).

As my other chapters show, Bose was part of a larger movement, which I term "*bhadralok* physics". This included other Indian scientists, such as Meghnad Saha and C.V. Raman, who espoused various forms of patriotism in different measures that were not incompatible with a *Visvajaneen* attitude in that they went beyond the contours of cultural distinctiveness.[13] Bengali linguist Suniti Chatterjee, a friend of Bose's, remarked:

> Satyendranath is convinced that the highest education in science in any country could and should be given through the medium of the mother-tongue ... I will also add that Bose does not have a segregationist mentality – he is not like those who would remove English from the Indian scene; as a practical man of science he will go for bilingualism for our higher scientific education, so long as the Indian scientists do not feel sure of themselves in their mother-tongues. But he would like the greatest support to be given to the mother tongue.[14]

Suniti Chatterjee's remark shows the extent of the *visvajaneen* cosmopolitan dimension in Bose's methodology of doing science. Bose espoused elements that were neither "British" nor "Indian" in a local sense but, in actuality, belonged to a wider transnational perception.[15] Arguing that Bose's methods were not "Indian" means that Bose did not conform to the prevalent educational system, in which there was no sustained effort in science education in the vernacular (e.g., Bengali). Bose's contributions to quantum physics show the different nature of his science. Rather than the "hybrid" framework sought out by the British administrator Thomas Babington Macaulay (in the 1830s), the newly emerging *bhadralok* intellectuals, such as Bose, developed a cosmopolitan, *visvajaneen* outlook, and were not necessarily looking toward Britain or India for examples or models. Scientists can further see this new direction through the lens of Bose's physics.

Bose's physics

Bose's first major contribution to theoretical physics was a paper he wrote with Saha, titled "On the Influence of the Finite Volume of Molecules on the Equation of State," and published in the *Philosophical Magazine*. An eventful year for the history of physics, 1919 had been an even more eventful year for Bose. He published two papers in the *Calcutta Mathematical Society* (founded by Prasad), as well as another paper he wrote with Saha called "On the Equation of the State," which appeared in the *Philosophical Magazine*.[16]

That same year, 1919, had also been very productive in physics, since Eddington had confirmed the predictions of Einstein's relativity theory, thereby transforming Einstein into an iconic figure in physics. *The Statesman*, the national daily newspaper published in Calcutta, sent a reporter to the astronomical observatory at the Science College to obtain an explanation, and a cabled confirmation of Einstein's prediction. Saha immediately wrote a popular exposition, and gave it to the reporter.[17]

Being attracted to the nuts and bolts of Einstein's works, Bose and Saha published a collection of papers on relativity. As Mehra remarks "Saha translated (to English) Einstein's 1905 paper on special relativity and Minkowski's 1908 paper on the fundamental equations of electromagnetic phenomena in moving bodies. Meanwhile, Bose translated (to English) Einstein's 1916 paper on the foundation of the general relativity,"[18] which came out as *Principles of Relativity* published by the University of Calcutta press in 1919. Though Einstein had given publishing rights to Methuen, who had wanted to discontinue the distribution of Bose and Saha's English translations, Einstein said that as long as the book remained in circulation in India, he had no objection.[19] Bose and Saha's translations were also the first English translations of Einstein's papers, which was quite a remarkable achievement, considering the fact that they were working in a colony far away from a European metropole.

Soon after translating Einstein's papers and having read Bohr's papers in the *Philosophical Magazine* on the correspondence principle,[20] Bose obtained Sommerfeld's papers from D.M. Bose on multiple quantization, and the fine structure of spectral lines.[21] In 1920, he published his paper, "On the Deduction of Rydberg's Law from the Quantum Theory of Spectral Emission," in the *Philosophical Magazine*.[22] Hence, knowledge generation in the British colony of India continued in a seamless fashion as seen through Bose's early career.

Years in Dhaka

Bose spent many years at what was then called the Dacca University, which came into being in 1921, as a direct consequence of the partition of Bengal in 1905. Dacca University is said to be a compensation given to Muslims in exchange for the nullification of the Partition of Bengal, which created a new province of East Bengal, with Muslims making up the majority of the population. The Muslims of East Bengal welcomed the partition of Bengal in 1905. They hoped that the creation of a new province would give them opportunities to develop and grow. In pre-partition days, most of the colleges were located in or around Calcutta. Out of the forty-five colleges, only fifteen were in East Bengal and Assam.[23] Partition had been nullified in 1911 by the British Government, and as a consequence, the Muslim leaders of East Bengal appealed to the Viceroy for a remedy in the form of a university. As a result, the government recommended the establishment of a university at Dacca.

The establishment of Dacca University as a new model university had been an achievement attributed to its vice-chancellor and teachers. Bose received an offer from the Dacca University for a readership in 1921 from the first vice-chancellor, Sir Philip Joseph Hartog.

Hartog asked Bose to send all his papers to Hartog along with his bio-data. Mr. Walter Jenkins, a professor of physics, sent his recommendations to Hartog in favor of Bose for this readership, specifically in physics. Jenkins expressed his opinion by saying that the most suitable appointment would be that of Bose or Saha (after examining all the applicants).[24]

Hartog considered both Saha and Bose as extraordinary researchers, and highly praised Bose for his paper on a problem of Statistical Mechanics that Bose published with Saha in the *Philosophical Magazine* in 1918. Jenkins (see Figure 3.2) also pointed out that, on the basis of information gathered from other academics, he found Bose a "very enthusiastic and talented person and an effort should be made to secure his services."[25]

Typically, South Asian historians have viewed the two centuries of British rule as that of conquest, deception, and hegemony.[27] In the context of science, and especially through Bose's life, we see the emergence of a different picture. There was a dialogue between the colonial intellectuals and Indian scientists, leading to the benefit of both British and Indian scientists. The dialogue brings to light the complex processes of intercultural negotiation and collaboration involved in the making of scientific knowledge. Jenkins and Hartog were the biggest supporters of Indian science. Without their support, there would not have been a Satyen Bose, or his cohort of scientists like Meghnad Saha or C.V. Raman.

In 1921, Bose moved to Dacca University and started teaching physics. He advised his students to:

> Never accept an idea as long as you are yourself not satisfied with its consistency and the logical structure upon which the concepts are based. Study the masters. These are the people who have made significant contributions to the subject. Lesser authorities cleverly bypass the difficult points.[28]

Figure 3.2 Walter Jenkins.[26]

Though initially occupied with relativity, Bose did not neglect statistical mechanics. He recalled reading, among other papers, Planck's *"Vorlesungen über die Theorie der Wärmestrahlung"* (Lectures on the Theory of Thermal Radiation) and *"Vorlesungen über Thermodynamik"* (Lectures on Thermodynamics), Boltzmann's *"Vorlesungen über Gastheorie"* (Lectures on Gas Theory), and Gibb's "Elementary Principles in Statistical Mechanics." Bose was not entirely happy in Dacca at first, and wrote to Saha about his grievances (see Appendix A.2 for letters in Bengali from S.N. Bose and Jnan Chandra Ghosh to Saha).[29]

Satyendranath Bose's letter is similar to Jnan Ghosh's letter to Saha, as both writers complained about the difficulties of being a scientist in a colony. The career trajectories of such scientists, who were far from the European metropole, did not follow the royal road to science. There were several difficulties, including a dearth of scientific journals and a lack of scientific apparatus, logistics, financial aid, bureaucratic support, as well as problems with the inability to be geographically mobile, and to travel to the centers of physics like Munich, Gottingen, Berlin, Copenhagen or London. Difficulties existed, but these difficulties did not deter scientists like Bose from their research. Bose's determination is obvious in his letter to Saha, written in 1921. This letter captures the spirit of the Raj.

Although the letter was written in Bengali, there are a few of English words in it; for example, the letter starts with "my dear Meghnad," and then we see the words "boycott," "graphic account," "correct," "Nicol lens," "eyepiece," "apparatus," "research," "journal," University," "delegate," and "science library" throughout the rest of the letter. It can be inferred that Bose did not have a segregationist mentality, nor a vision that espoused only the local; but rather he had a broader vision that was open to learning from a wider global scene, and using this knowledge in everyday life. Though Bose would become one of the leading innovators in Indian science in the mother tongue, one can appreciate the *visvajaneen* aspect of Bose's thought process in his letter. (See Appendix A.2 for the original letter of Jnan Ghosh and Satyendranath Bose in Bengali, and see below for the English translation of the letter by the author).

Indian chemist Jnan Ghosh's letter to Saha (about Bose, and a scientist's life under the Raj) (Translated from Bengali):

The Chummery[30], Ramna, Dacca, 19.7.21,

My dear Meghnad, It has been a long time since I heard from you. After coming here, I am going through a lot of trouble. Calcutta University has laid me off after a while, but Dacca University is still giving me some trouble regarding my pay. Hartog is really interested to keep me in Dacca, as I really get along with Watson. Probably my payroll- troubles will end quickly and I can get back to research. As a representative of Calcutta University I have received a lot of invitations to present my research. I am interested to know more about your present research. But the research in Calcutta University is getting better.

I sent a proof sheet of my paper to *Physical Chemistry B*. In the coming weeks I will correct my paper and I think it is better to stay abroad than coming back to your own country as the research conditions there are better. When are you coming back? Otherwise, everything is fine. Satyen (Bose) is now Dacca University's darling and works really hard in his research.

Yours affectionately Jnan[31]

Bose's letter to Saha (translated from Bengali):

19.7.21,

My dear Meghnad,

I have not received any letters from you but very often in Calcutta I used to hear from you. I believe as I have been irregular in writing letters to you, you can boycott me. Jnan and myself are living at the same place. I heard from Jnan that you have visited Germany and met several stalwarts of physics and you were supposed to visit Munich. I can expect a graphic account from you about your international research experiences.

Over a month now I have come to your hometown. The work here has not yet begun. In your Dacca College there was lots of research equipment, but due to lack of maintenance, they are in a terrible state. Perhaps you have some idea about it. There is a lot of experimental apparatus like Nicol prisms, lens, eyepiece scattered all over the tables but one has to do some research to ascertain which parts belong to what apparatus. Experimental research would be possible if we can fix the lacuna in apparatus acquisition and a good lab.

There is a dearth of journals here, but there is talk of a new University at Dacca and then I hear the British members would order a lot of new journals as per their assurance and also a separate science library. That is all that is going on here. Now that you are a delegate of Calcutta University, What did you see? What did you do over there? Have you received any honorary degree?

Satyen[32]

Learning physics by reading books not written in English would make the process of understanding the basic concepts more difficult. Thus, researching as a scientist under imperial rule used to be difficult but not impossible, as Bose's career shows; and it also shows how, in certain cases, the *bhadralok* scientists grappled with difficulties and achieved success in research, by publishing papers and networking with scientists locally and globally.

In 1923, Bose was teaching Thermodynamics and Electromagnetic Theory to the M.S. classes at Dacca University.[33] He studied the Theory of Relativity simultaneously with Quantum Theory. While working on Quantum Theory, Bose felt the need for a logically more satisfactory derivation of Planck's Law. The apparent problem had been in the derivations of Planck's radiation Law, which gives the energy distribution of electromagnetic radiation (also known as black body radiation) in equilibrium at temperature T as a function of frequency v.

$$\rho(v,T) = 8\frac{\pi v^2}{c^3}\left(\frac{hv}{e^{\frac{hv}{kT}}-1}\right)$$

In the equation, ρ is the energy density per unit frequency interval at frequency v and temperature T.[34] This equation has been written as a product of two factors, each factor playing a different role in derivations of the formula.

Planck's law, insofar as he conceptualized it, had been just a lucky guess in keeping with experimental data; the theoretical foundations were slightly shakier. Planck, in a colloquial fashion, remarked later to the American physicist Robert W. Wood (1931), that the theoretical underpinning of his radiation law was a high-stakes concept, which ultimately needed to be found. The photon concept, as Thomas Kuhn convincingly argues, arose in 1905, championed by Einstein.[35]

Early-twentieth-century physicists were also grappling with the dual nature of light. Though Europe was bound in the tentacles of Maxwellian electromagnetic continuum, i.e., the wave nature of light, Einstein chose to conceptualize light differently. Einstein introduced the light quantum hypothesis in 1905, showing that light is a particle. Later, in Salzburg in 1909, Einstein introduced his fluctuation formula, which showed that the mean square fluctuation in energy was the sum of a linear term and a square term in radiation energy density, representing the particle and wave terms. According to Martin Klein, "Einstein concluded that there were two independent causes producing the fluctuations and that an adequate theory of radiation would have to provide both wave and particle mechanisms."[36] This conclusion laid the seeds of the puzzling wave-particle duality and complementarity, which would later create an unsettling discourse in the foundations of quantum physics.

Though it was increasingly suggestive that radiation inside a cavity should be thought of as a "photon (quantum) gas", there were several reasons why Einstein did not pursue a "quantum gas" model of blackbody radiation around 1905. Light quanta's only particle-like characteristic was energy. More importantly, Einstein was well aware that light quanta could not be statistically independent of each other, as regular classical particles were; and if they were statistically independent, a "photon gas" would obey Wien's law and not Planck's. Einstein remarked:

> Indeed, I am not at all of the opinion that one should think of light as composed of quanta localized in relatively small spaces and independent of each other. This would indeed be the most convenient explanation of the Wien end of the radiation formula. But by itself, the division of a light ray at the surface of the refracting medium completely forbids this conception. A light ray can divide, but a light quantum cannot divide without change in frequency.[37]

Einstein also did not pay much attention to statistics. His quirky approach towards probability, his European education, and his 1909 Salzburg wave and particle fluctuation formula, which eventually laid the seeds of wave-particle duality, made his modus operandi different from that of Bose who had none of those all-encompassing features. The discourse about the puzzling wave-particle duality and skepticism towards the discontinuous nature of light was, however, unknown to Bose, who was a scientist in a colony living far from the European metropole, with no established traditions of classical physics.

Following the observations of Paul Ehrenfest (1906), that the energy of field excitations should be quantized, Peter Debye (1910) re-derived Planck's law using the notion of quantization of elastic vibrations, to account for the specific heat of solids. Einstein (1916) gave yet another phenomenological derivation of Planck's law based on radiative equilibrium, resulting from the simultaneous consideration of stimulated and spontaneous emission, and the introduction of his famous A and B coefficients.

Small wonder then, that Bose remarked in his first 1924 paper, "In every case, the derivations do not appear to me to be sufficiently logically justified,"[38] because none of the previous derivations, including those by Planck (1900), Peter Debye (1910), Albert Einstein (1916), Paul Ehrenfest (1923) and Wolfgang Pauli (1923), "were based on consistent use of quantum concepts. Each involved classically-derived results along with new quantum ideas."[39] Bose continued his assessment of quantum concepts when he said:

> As a teacher who had to make these things clear to his students I was aware of the conflicts involved and had thought about them. I wanted to know how to grapple with the difficulty in my own way. It was not some teacher who asked me to go and solve this little problem. I wanted to know. And that led me to apply statistics.[40]

Among the physics literature available to Bose at that time was Peter Debye's 1910 derivation of Planck's Law, which was published in *Annalen der Physik*. In the introduction of Bose's second 1924 paper, "Thermal Equilibrium in Radiation Field in the Presence of Matter," he gave a critical review of the existing papers that dealt with important derivations of Planck's Law, and he remarked:

> Debye has shown that Planck's law can be derived with the aid of statistical mechanics. However, his derivation is not completely independent of classical electrodynamics insofar as he makes use of the concept of proper vibrations of the ether [that is, normal modes of the field] and assumes that with respect to energy the spectral region between ν and $(\nu + d\nu)$ can be replaced by $(8\pi\nu^2/c^3)$ V dν oscillators, whose energy can only consist of multiples of hν. However, one can show that the derivation can be altered so that one does not have to borrow anything from classical theory.[41]

As we see, Bose's analysis of Debye's derivation, along with the derivations of Planck, Ehrenfest, Einstein, and Pauli, revealed a medley of classical and quantum concepts that were difficult to teach in a seamless fashion. This jumble of concepts was the primary motivation for Bose to derive Planck's Law, which was based on the energy quantum hypothesis. Using statistics as his thrust, Bose derived the two factors in Planck's Law, completely independent of classical electrodynamics, using particle-like entities called quanta that can only have discrete energies.

Bose relied on phase-space arguments, treating radiation inside a cavity as an ideal photon gas, with each photon having energy hv and momentum (hv)/c. Though energy and momentum are properties of particles, and many physicists in Europe would be skeptical about attributing particle-like properties to light, Bose had no such prejudices. The frequency distribution of radiation at an absolute temperature T is then deduced by finding the distribution in phase space by maximizing the entropy of the system, which is the criterion for equilibrium of radiation. As light-quanta (treated as a photon gas by Bose) are dealt with unlike non-interacting atoms, their number is not conserved. They are mass-less, hence they should be treated relativistically. Most importantly, the photons were treated by Bose as *indistinguishable*. This property of being indistinguishable or identical was a novel contribution by Bose. His attention to statistics paid off here, because in classical statistics these phase points would be regarded as distinct. As a consequence, Bose's first paper, "Planck's Gesetz und Lichtquantenhypothese" re-derived Planck's law from purely non-classical considerations.[42] It was a derivation that leading European scientists, including Einstein, had failed to achieve. Hence this can be seen as a special moment where knowledge is being generated from a colony, and later being used by metropolitan scientists for furthering the development of physics on a global scale.

Bose sent the above paper to the *Philosophical Magazine* in 1923. When he did not hear from them, he sent a copy to Einstein for his opinion, with a request to have it translated into German, and published in the German journal, *Zeitschrift für Physik*. It is still not known why there was no response from *Philosophical Magazine*. One might deduce from this inaction that the paper was not rejected, but merely ignored. Presumably, Einstein was very much pleased by his paper, and that is why he immediately sent a letter of praise to Bose, calling his work a beautiful step forward.[43] Bose's letter to Einstein, dated June 4, 1924, began as follows:

> Respected Sir, I have ventured to send you the accompanying article for your perusal. I am anxious to know what you think of it. You will see that I have ventured to deduce the coefficient $(8\pi v^2/c^3)$ in Planck's Law independent of classical electrodynamics, only assuming that the ultimate elementary regions in the phase-space has the content h^3.[44]

It took some time for Einstein to digest and translate Bose's work, and to conceive and complete his own extension of it with regard to the quantum ideal gas. We should analyze why Einstein became thoroughly convinced of Bose's precision.

To understand Einstein's surety, one must go back to Einstein's "photoelectric effect" paper of 1905.[45] His "light quantum hypothesis" undermined the continuum structure of Maxwellian electrodynamics.

In a lecture at the 1909 Salzburg conference, Einstein famously prophesied "that the next phase of the development of theoretical physics will bring us a theory of light that can be interpreted as a kind of fusion of the wave and emission theories."[46] "Contrary to what the term "fusion" [*Verschmelzung*] in this quotation suggests," as Michel Jannsen and Anthony Duncan argue, "Einstein believed that his 1909 fluctuation formulae called for two separate mechanisms: the effects of the two causes of fluctuation mentioned [waves and particles] act like fluctuations(errors) arising from mutually independent causes."[47]

Nevertheless, as Alexei Kojevnikov argues,

> Einstein did not use the word "duality" either before or after 1925, nor did he make any clear assertion of the principle of the wave-particle duality. Einstein accepted the wave-particle duality only in a negative sense, as a fundamental difficulty of quantum theory that had to be resolved rather than turned into a postulate.[48]

Bose's derivation gave Einstein that answer, as it vindicated a view he had championed for roughly nineteen years. Conversely, it was not just a "shot in the dark" that Bose triggered, as Pais argues.[49] He was very much aware that his result was a logical development of Einstein's work, which had eluded Einstein for nearly two decades. Consequently, Bose knew who to appeal to, but was also isolated enough not to reject Einstein's outlook.[50] Bose's first letter (June 4) to Einstein from Dacca University in 1924 (as mentioned earlier in part), goes on to say:

> I do not know sufficient German to translate the paper. If you think the paper worth publication, I shall be grateful if you arrange for its publication in *Zeitschrift für Physik*. Though a complete stranger to you, I do not feel any hesitation in making such a request. Because we are all your pupils though profiting only by your teaching through your writings. I do not know whether you still remember that somebody from Calcutta asked your permission to translate your papers on Relativity in English. You acceded to the request. The book has since been published. I was the one who translated your paper on Generalized Relativity.
>
> Yours faithfully,
> S.N. Bose[51]

A few days later on June 15, he wrote to Einstein:

> Respected Master,
>
> I send herewith another paper of mine for your kind perusal and opinion. I hope my first paper has reached your hands. The result to which I have

arrived seems rather important (to me at any rate). You will see that I have dealt with the problem of thermal equilibrium between Radiation and Matter in a different way, and have arrived at a different law for the probability for elementary processes, which seems to have simplicity in its favour. I have ventured to send you the type-written paper in English. It being beyond me to express myself in German (which will be intelligible to you), I shall be glad if its publication in *Zeitschrift für Physik* or any other German journal can be managed. I myself know not how to manage it. In any case, I shall be grateful if you express your opinion on the papers and send it to me at the above address.

Yours truly,
S.N. Bose[52]

From Paris, Bose wrote to Einstein in 1924:

17 Rue de Sommerard
Paris
26th October 1924

Dear Master,

My heartfelt gratitude for taking the trouble of translating the paper yourself and publishing it. I just saw it in print before I left India. I have also sent you about the middle of June a second paper entitled "Thermal Equilibrium in the Radiation Field in the Presence of Matter.' I am rather anxious to know your opinion about it, as I think it to be rather important. I don't know whether it will be possible also to have this paper published in *Zeitschrift für Physik*. I have been granted leave by my university for 2 years. I have arrived just a week ago in Paris. I don't know whether it will be possible for me to work under you in Germany. I shall be glad, however, if you will grant me the permission to work under you, for it will mean for me the realization of a long-cherished hope. I shall wait for your decision as well as your opinion of my second paper here in Paris. If the second paper has not reached you by any chance, please let me know. I shall send you the copy I have with me.

With respects,

Yours sincerely,
S.N. Bose[53]

On July 2, 1924, Einstein sent a postcard to Bose[54] (see Appendix A.5): "Dear Colleague, I have translated your paper and given it to the *Zeitschrift fur Physik* for publication. It signifies an important step forward and pleases me very much."[55]

On July 12, 1924, Einstein wrote to Paul Ehrenfest: "The Indian Bose has given a beautiful derivation of Planck's Law, including the constant ($8\pi v^2/c^3$)."[56]

Einstein also added the comment: "the derivation is elegant but the essence remains obscure."[57]

To understand Einstein's last comment, we need to ponder over Bose's perspective. In re-deriving Planck's Law, Bose had introduced a coarse-grained counting method to count the number of states in a certain frequency interval. But instead of counting wave frequencies, Bose counted cells in one particle phase space, and divided the resulting expression by the volume of a cell, and multiplied the resulting expression by "2" to take "polarization" into account. As Bose's last graduate student, Partha Ghose, argues, "no explanation is offered as to how this 'polarization,' an essentially classical concept, can be understood in terms of the light-quantum hypothesis; although Bose claimed to deduce it independent of classical electrodynamics."[58] So it understandably remained obscure to Einstein in 1924.

Further, as Ghose remarks, "Bose had always maintained privately that he did offer a quantum theoretic explanation, but Einstein removed it from his translation and substituted it with the statement about the polarization factor '2'."[59] In his letter to Bose on July 2, 1924, Einstein wrote, "You are the first person to derive the first factor theoretically, even though not wholly rigorously."[60]

Bose's explanation was

> that light-quanta carried an intrinsic spin that could take only the values "±h/2π". There is no recorded evidence of this because Bose's original manuscript in English is missing from the Einstein archives. It appears in a paper by C.V. Raman and S. Bhagavantam (1931) entitled "Experimental Proof of the Photon Spin."[61]

In their paper (reproduced verbatim), Raman and Bhagavantam wrote:

> In his well-known derivation of the Planck radiation formula from quantum statistics, Prof. S.N. Bose obtained an expression for the number of cells in phase space occupied by the radiation, and found himself obliged to multiply it by a numerical factor 2 in order to derive from it the correct number of possible arrangements of the quantum in unit volume. The paper as published did not contain a detailed discussion of the necessity for the introduction of this factor, but we understand from a personal communication by Prof. Bose that he envisaged the possibility of the quantum possessing besides energy hν and momentum hν/c also an intrinsic spin angular momentum ±h/2π round an axis parallel to the direction of its motion. The weight factor 2 thus arises from the possibility of the spin of the quantum being right-handed or left-handed, corresponding to the two alternative signs of the angular momentum. There is a fundamental difference between this idea, and the well-known result of classical electrodynamics to which attention was drawn by pointing and more fully developed by Abraham that a beam of light may in certain circumstances possess angular momentum associated with a quantum of energy is not uniquely defined, while according to the view we are concerned with

in the present paper, the photon has always an angular momentum having a definite numerical value of a Bohr unit with one or other of the two possible alternative signs.[62]

Raman and Bhagavantam, being experimentalists, described and discussed their observations in the *Indian Journal of Physics* in 1931, which had led them to the conclusion that the light quantum possessed an intrinsic spin equal to one Bohr unit of angular momentum. In a short letter to *Nature* in January,1932, they confirmed their conclusions with an improved apparatus. Their experiment determined the extent to which the depolarization of Rayleigh scattering of monochromatic light is diminished, when it is spectroscopically separated from the scattering of altered frequency arising from molecular rotation in a fluid.[63] It can be argued that the concept of the photon spin, which, has been established without doubt in present day physics, was proposed by Bose in 1924.

On June 15, 1924, Bose sent his second paper to Einstein entitled "Thermal Equilibrium in the Radiation Field in the Presence of Matter."[64] Using Einstein's postcard as a reference, Bose was able to get a quick approval for a study leave to go to Europe. In October of 1924, he arrived in Paris, thinking he would spend a few weeks there before going to Berlin for visiting Einstein. Because he was more comfortable in French than in German, he thought that language would not be a barrier in Paris. He met Madame Curie and wanted to learn more about radioactivity in her lab.[65] Madame Curie was of the view that people working in her lab should know French very well, so she told Bose, "Why don't you go and learn some French first, and then report here?"[66]

Bose was too shy to tell her that he had been studying French for the past fifteen years. Instead, he told her that he could stay in Paris only for six months. As he was very polite (*bhadra*), he could not tell her that he was more interested in physics than in French. In reply, Madame Curie cautioned him not to hurry, and advised him to concentrate on the language first. After this incident, Bose lost interest in working with Madame Curie, However, it did not deter his overall scientific interests. Through Paul Langevin, he came into contact with the de Broglie brothers—Maurice de Broglie, from whom he learned many techniques in X-ray spectroscopy and X-ray crystallography; and Louis de Broglie, who had just submitted his revolutionary thesis to the Sorbonne[67] (Figure 3.3).

Bose wrote a paper entitled "Fluctuations in Density," which ends with a fundamental result showing that, based on the new counting method initiated by him, the mean square energy fluctuation of the gas molecules is given by an expression which is the sum of two terms. The first term (proportional to frequency) corresponds with the Maxwell–Boltzmann statistics of non-interacting molecules, and the other term (proportional to square of frequency), because of interference fluctuations, is associated with wave phenomena.[69]

From this calculation (as Partha Ghose recollects from his conversations with his mentor), "Bose drew attention to the importance of Louis de Broglie's doctoral thesis that he had heard of from Paul Langevin"[70] in Paris. He "asked for a copy which he received and read in December of 1924 and found that Louis

Figure 3.3 Bose in Paris with Bertrand Zadoc Kahn.[68]

de Broglie had attached wave properties to matter in analogy with wave-particle duality of radiation."[71] On this explanation, Einstein remarked: "I shall discuss this interpretation in greater detail because I believe that it involves more than a mere analogy."[72]

Partha Ghose argues that indistinguishability was a novel idea, which was introduced by his mentor Bose through his statistics. Additionally, the formalism of quantum mechanics called "wave-mechanics," as pioneered by Erwin Schrödinger, was very much dependent on Bose's idea of indistinguishability.[73] Therefore, Bose's contribution to the unfolding field of quantum mechanics in the mid-1920s cannot be disregarded, or taken as a random shot in the dark (Figure 3.4).

John Stachel remarks about the Einstein–Bose interaction in Berlin (1925), that Einstein had proposed two problems to Bose to work on:

first the question of whether the new statistics implied a novel type of interaction between the light-quanta, and second what the statistics of light-quanta and transition probabilities of radiation would look like in the new theory (i.e. quantum mechanics). Apparently, Bose made no progress on either problem.[75]

Figure 3.4 Bose with a friend in Berlin.[74]

Focusing on the first problem, Bose's statistics implied certain strange quantum correlations between distant particles; and this idea intrigued Einstein. This interest is the reason why Einstein asked Bose to look at the type of novel interactions the statistics implied in the new theory. However, as we know today, the idea of "entanglement" does not suggest that these correlations imply any interactions at all in the new theory. Hence Bose understandably did not find any answer, if at all he tried to find one in Berlin.

Moving on to the second problem, Bose disagreed with Einstein on the nature of transition probabilities (Appendix A.7). Bose's second paper (1924), "Thermal Equilibrium in Radiation Field in the Presence of Matter," as Ghose states,[76]

> derived general conditions for statistical equilibrium of a system consisting of matter and radiation, independent of any special assumptions about the mechanism of the elementary radiative processes. In the second part of the paper, Bose proposed a new expression for the probability of these elementary radiative processes that differed from what Einstein[77]

proposed in 1916. Bose's theoretical result consisted of finding the probability of an interaction P, which was calculated as:

$$P = (Ns\, dv_s)/(A_s + Ns\, dv_s) = (\check{n})/(\check{n}+1) \text{ where } \check{n} = N_v/A_v$$

This interaction represents the average number of quanta per cell. v is the frequency, A is the spontaneous emission constant. Ns is the number of light quanta of energy hv_s.

Bose's probability law predicted a dependence of the absorption coefficient on the radiation density, decreasing with the radiation density. The departure from classical behavior predicted by Bose's principle should occur only when ň $\ll$ 1, i.e. for very low intensity radiation.[78]

Though such light sources were not experimentally available in 1924, with the advent of modern-day quantum optics there are states, "like single photon states (Fock states) and squeezed states"[79] showing sub-Poissonian statistics, which are a feature of non-classical physics.[80]

Consequently, Bose acknowledged that the division between spontaneous and stimulated emission as two independent processes did not appear natural, and was quite unnecessary. This was possibly the reason why Bose did not work out the ramifications of the new theory, even if he was really asked to do so by Einstein.[81] Ghose claims that:

> It is puzzling that even after championing Bose's method, Einstein failed to see Bose's point about spontaneous emission. To Bose it was clear that the cause of spontaneous emission depended on the environment in which the atom was placed, which Einstein did not approve of. It is well known presently that the cause of spontaneous emission is vacuum fluctuations as later put forward mathematically by Paul Dirac's second quantization.[82]

Bose's critique of Einstein's 1917 paper, "Stimulated and Spontaneous Radiation," in Bose's second and third papers, led him to introduce photon spin, and his statistical understanding of how to conceptualize an electromagnetic field.[83] His second paper, on the interaction of matter and radiation, was much longer than his first one, and seemingly more ambitious. He rejected Einstein's special assumptions about there being two kinds of radiative processes, *spontaneous and induced,* by means of which an atom of higher energy makes a transition to a lower energy level. Einstein's 1917 proof of Planck's law had been constructed phenomenologically, in a somewhat contrived manner. Bose then claimed that the transition from the higher energy state to the lower energy state can be explained more elegantly, without bringing in Einstein's (additional) hypothesis of an induced transition.

Furthermore, Bose claimed that spontaneous transition was enough to explain the transition from a higher energy state to a lower energy state, which could in turn be understood as a property arising from the statistical character of the radiation field itself, consistent with all the equilibrium conditions. Thus, the emission of light could be viewed as a unified single process—arising purely from the statistical property of the radiation field itself—and not something that was dependent upon the specific causal mechanisms of energy transfer. This conclusion was analogous to Bose's own earlier conclusion in the first paper, where he had derived Planck's law by essentially arguing in the same manner.

Einstein had strong objections to these proposals. In his note to Bose on November 3, 1924, Einstein objected that the absorption coefficient is independent of the radiation density, as confirmed by experimental evidence from infra-red radiation. In response to this objection, Bose wrote a third paper, and sent it to Einstein from Paris.

He forwarded the third paper to Einstein on January 27, 1925, during his stay in Paris[84] under a separate cover. It was accompanied by a letter[85] to Einstein with the information that Langevin had found the paper interesting and worth publishing. The letter stated that Bose "tried to look at the radiation field from a new standpoint and have sought to separate the propagation of Quantum of energy from the propagation of electromagnetic influence."[86] The communication also included a mention of the Bohr–Kramers–Slater (BKS) theory (1924) and its method of "virtual oscillators," which Bose was aware of and had found similar to his treatment of the spontaneous emission. While Bohr was a radical supporter of the wave theory, Bose was quite comfortable using light quanta.

Unfortunately, the text of the third paper presented by Bose to Einstein still remains untraceable. The contents of the third paper, therefore, remain unknown, since Einstein did not send it anywhere for publication, nor did he return it to Bose,[87] because the latter discussed his new ideas personally with Einstein in October 1925 in Berlin. Einstein, however, could not accept Bose's explanations of the contents of the third paper. Bose most likely abandoned the draft after hearing Einstein's further critical responses to spontaneous and induced radiation, which, Einstein maintained, were two very distinct processes. Heartbroken, Bose returned to India in 1926, concentrating on teaching and guiding researchers in various scientific pursuits.[88]

Einstein's apparent doubts about some of Bose's ideas reveal the complexities about the stages of development of science in British India. It also shows how Bose participated in the making of new scientific knowledge, and how Indian scientists took their first steps toward creating an indigenous modernity, independent of colonial connotations. If Bose's recognition abroad is important for Indian science, his frustration regarding his second and third papers is equally significant for an insightful understanding of science in colonial India. Bose made no further attempt to publish his paper, even though his note to Einstein mentions that he had shown it to the French physicist Paul Langevin, who had thought it worthy of publication.[89]

In what sense were Bose's research and work in physics "modern"? First, "modernity" in physics was essentially related to a kind of representation that was deemed acceptable.[90] The onset of modernity in Bose's physics was associated with a shift from a mechanical, continuous, and envisioned, to an abstract and non-intuitive discontinuous representation of the light quantum's properties and behavior. Bose distanced himself from the British mechanical style of concrete model-building, and embraced the continental abstract style of theorizing. Thus, the modernist style of Bose's physics did not appear out of nowhere.

Bose's scientific approach was such because he was influenced by circumstances to abandon mechanical representations and the continuum nature of light. The intellectual milieu in which he was raised did not have a tradition of the classical-continuum physics of Maxwell, as in Europe in the late nineteenth and early twentieth centuries. In his early education in Bengal as well, Bose was influenced by the nature of historical writing, especially by his mentors Jagadish Chandra Bose (JCB) and Prafulla Chandra Ray (PCR). JCB and PCR espoused a worldview which conceptualized Indian history in terms of ruptures and discontinuities. First, there was a classical Indian era of native kings in South Indian Dravidians that was halted in its progress by invasions, first from the North-West, most notably the Aryans, then the Mughals, and lastly the British. There was a manifest rupture, geographically, ideologically, and administratively, from the Partition of Bengal, which also predisposed Bose's worldview in science toward discontinuity.

Bose always referred to Einstein in his letters as "Respected Master." Einstein had become, in Bose's worldview, his intellectual *guru*. The *Guru–Shishya* relationship has a long history going back to the Indian epic, the *Mahabharata*. The Indian epic upholds the sublime nature of the *Guru–Shishya* relationship. As Einstein did not share a similar history, he conceptualized Bose as just another scientist and colleague, not a *shishya* (pupil). As an Indian, Bose viewed Einstein with great reverence, following the tradition of the *Mahabharata*. He viewed the rejection of his third paper by Einstein as nullification from his revered *guru*. Consequently, he did not proceed further with the paper, in spite of the approval of Langevin. This occurrence can be viewed as a loss to the development of physics on a global scale.

Quite a few biographies about Bose belong to the hagiographic genre, which blindly praise him for his scientific accomplishments. This is not atypical in either the history of science or South Asian history. Considering that Satyendranath Bose is not very well known globally, these hagiographies (as mentioned earlier) are better than a total absence of historical narratives. This chapter, however, has a dual role. First, in a non-hagiographical way it clarifies Bose's scientific contributions and motivations for deriving Planck's Law. Second, this chapter shows that physics does not operate in a cultural vacuum. Bose was a non-Western scientist who worked under British rule in a power differential, and successfully negotiated opportunities that came his way when he was an early career scholar with no permanent institutional base.

In spite of being from a colonized country, Bose showed mastery over cutting-edge physics—quantum mechanics—which was slowly unfolding in various contradictory ways in Europe. This contribution is where Bose deserves credit, because his ideas led to the conceptual development of quantum mechanics through his concept of indistinguishability or identity of particles, characteristic of quantum particles, which ultimately led to the development of wave mechanics by Erwin Schrödinger.

Older Eurocentric histories of science, like the ones by George Basalla, have revealed how scientists working in various colonies were passive recipients of Western scientific modernity.[91] Through Bose's case study, we see that such linear diffusion of scientific modernity did not happen. It was far more complex than that, as, through the lens of Bose's early life, we see the peculiar entanglement between the local and the transnational. Bose's attraction toward German physics as an escape from the colonial situation, and his upbringing as a student during the Partition of Bengal—marked by a manifest geographical discontinuity—clearly played a role in determining Bose's physics. His innovations in physics showed an affinity for quantum discontinuity. He also took Einstein's light quantum seriously, unlike any other scientist in Europe, which was unusual for an early-career intellectual working in a colony.

Bose's European sojourn, and his interactions with European metropolitan scientists—Marie Curie, Paul Langevin, Maurice de Broglie in France, and Albert Einstein—all added up to produce a very transnational nature of his physics. The global outreach of this physics was peculiar, because it was also rooted in the nation, and the Indian political and social context. This can be taken as an example of how knowledge circulates, and is co-produced transnationally for the benefit of a global science.

I also want to point out here that Bose's inadvertent isolation from the physics community of Europe while working in India (before going to Europe) led him to devise his new statistics, because his isolation may have shielded him from the lingering skepticism about Einstein's work on the light quantum. Furthermore, Bose's thinking was very much like Einstein's, especially the motivation to re-derive Planck's Law, independent of classical physics.

Bose–Einstein statistics showed essentially the interconnected nature of Indian and German physics. Bose displayed through his science that there was no antagonism between the local and global; rather, one could complement the other. Hence, the collaboration between Bose and Einstein shows the locally rooted cosmopolitan nature of *bhadralok* physics and Indian modernity. Local *Visvajaneenata* cosmopolitanism did not simply mean looking outside the nation. Bose was also a nationalist. His return to India after the European sojourn, and his efforts to institutionalize science and train students are indicative of that reality.

Lastly, I propose a general framework in my concluding chapter, advising on how to engage with such multi-hybrid intellectual patterns of thought in which India, France, Germany, or any mix of nations are involved in co-producing a scientific idea. I hope my framework, as outlined in the conclusion, will be helpful to all forthcoming intellectual historians who engage with such hybrid mixing of scientific ideas. This approach, I believe, is completely new, and offers a far more nuanced way of dissecting Indian modernity using a multi-hybrid analysis, than the methodology displayed by South Asian historians[92] and cultural studies scholars.[93]

Notes

1 Somaditya Banerjee. "Transnational Quantum: Quantum Physics in India Through the Lens of Satyendranath Bose." *Physics in Perspective* 18, 2 (2016) 157–181.
2 In the chapter on Raman, I explain the nature of experimental work done by C.V. Raman and his collaborators on light scattering. But Raman started out in theory when he did not get access to proper labs and equipment and gradually shifted to theory and experiment when he became the Palit professor of physics at Calcutta University.

3 For example, the perception of radicality of quantum physics was maintained and propagated centrally by the 1911 Solvay Congress.

4 Credit: SOAS, University of London. I thank David Ogden of SOAS for this picture.

5 Jagadish Mehra and Helmut Rechenberg. *The Historical Development of Quantum Theory*. (New York: Springer 1982) 568–570.

6 www.snbose.org (January 24, 2013) This document has been edited by Bose's grandson Falguni Sarkar. The letter in English is reproduced here in its original unedited form including typos.

7 Ibid.

8 SNBCS Archive, Doc 0006.

9 www.snbose.org (accessed 15 Dec 2012).

10 Ibid. Now these letters are part of the SNBCS Digital Archive at Kolkata, India. I thank Prof. Samit Ray, Director of SNBCS for giving me access to these letters.

11 Ibid. The original document is being reproduced here verbatim. It is interesting that Eisenmann characterizes Meghnad Saha and Satyendranath Bose as "self-taught" which is only partially true as both Bose and Saha had mentors in college. But as the letter of Eisenmann is hagiographic, "self-taught" sounds more in line with a hero-like image.

12 Mahadev Dutta (student of Satyendranath Bose) in conversation with Meghnad Saha's student Santimay Chatterjee. See Santimay Chatterjee and Enakshi Chatterjee. *Satyendra Nath Bose*. (Calcutta: National Book Trust, 2005) 41.

13 For the case of Indian poet and Nobel Laureate Rabindranath Tagore, Harvard historian Sugata Bose has shown how a similar cosmopolitanism manifested as a different universalism. A universalist patriotism termed by Sugata Bose as "Cosmopolitan Thought Zones" emerged in several colonies. See Sugata Bose and Kris Manjapra. *Cosmopolitan Thought Zones: South Asia and the Global Circulation of Ideas*. (New York: Palgrave Macmillan, 2010) 97–111.

14 SNBCS Archives, Doc 0030.

15 For a similar argument in a different context see Claude Markovits. "How British was British India? Recovering the Cosmopolitan Dimension in the British-Indian Colonial Encounter." *Jahrbuch fur Europaische Uberseegeschichte* 10 (2010) 67–91.

16 Meghnad Saha and Satyendranath Bose. "On the Equation of State." *Phil. Mag.* 6, 39 (1920) 456.

17 Robert Anderson. *Nucleus and Nation: Scientists, International Networks and Power in India*. (Chicago: University of Chicago Press, 2010) 9.

18 Mehra. "Satyendranath Bose" 122.

19 H. A. Lorentz, A. Einstein, H. Minkowski, and H. Weyl. *The Principle of Relativity: A Collection of Original Memoirs on the Special and General Theory of Relativity* (with notes by A. Sommerfeld), trans. by W. Perrett and G.B. Jeffery (London: Methuen & Co., 1923).

20 N. Bohr, "On the Quantum Theory of Radiation and the Structure of the Atom." *Phil. Mag.* 30 (1915) 394–415.

21 A. Sommerfeld. "Zur Quantentheorie der Spektrallinien." *Annalen Phys.* 51, 1–94 (1916) 125–167; (Wlth W. Kossel) "Auswahlprinzip und verschiebungssatz bei serienspektren." *Verh. dt. phys. Ges.* 21 (1919) 240–259.

22 S.N. Bose. "On the Deduction of Rydberg's Law from the Quantum Theory of Spectral Emission." *Phil. Mag.* 40 (1920) 619.

23 Santimay Chatterjee and Enakshi Chatterjee. *Satyendra Nath Bose*. (Calcutta: National Book Trust, 2005). 17.

24 Ibid. 34.

25 Santimay. "Satyendra Nath Bose" 34–36.

26 I thank Professor Dr. Sabbir Ahmed, the Managing Editor of Banglapedia Trust, for giving me access to this photo.
   http://en.banglapedia.org/index.php?title=Jenkins,_Walter_Allen (accessed September 15, 2019).

27 Gyan Prakash. *Another Reason* 10–12, 83; Shashi Tharoor. *Inglorious Empire*: *What the British Did to India*. (London: Hurst & Company, 2017).

28 SNBCS Archives, File no. 275 (accessed January 10, 2012).

29 Oral interview with Partha Ghose with the author on January 16, 2012.

30 Incidentally, when I was researching at the archives in India, I was not familiar with the word "chummery." I asked the archivist (who was also a theoretical physicist) at Calcutta what it meant. He said it was not "chummery" but "chimney." I was perplexed why Bose wrote a letter from someone's chimney. When I pushed the archivist about the word, he agreed that I might be right, but he was unfamiliar with the term "chummery." On being pushed further, he mentioned that he was born much after 1921, so it is completely justified that he did not know much about it. I mention this episode just to give the reader a glimpse of working at Indian archives.

31 See Appendix A.2 for full letter.

32 See Appendix A.2 after Jnan's letter

33 Santimay Chatterjee and Enakshi Chatterjee, 40.

34 "c" is the speed of light, K is Boltzmann's Constant, and "h" is Planck's constant. See John Stachel. *Einstein from 'B' to 'Z'*. (Boston: Birkhauser, 2002) 519–538.

35 Here I agree with Thomas Kuhn who argues that Einstein introduced the quantum in 1905. Peter Galison. "Kuhn and Quantum Controversy." *The British Journal for the Philosophy of Science* 32, 1 (March 1981) 71–85.

36 Martin Klein. "The First Phase of the Bohr Einstein Dialogue." *Historical Studies in the Physical Sciences* 2 (1970) 6.

37 Einstein to Lorentz, 23 May 1909, Doc.163 as quoted in *The Collected Papers of Albert Einstein, vol.5, The Swiss Years: Correspondence, 1902–1914*, Martin J. Klein et al., eds. (Princeton: Princeton University Press) 193.

38 S.N. Bose. "Planck's Gesetz und Lichtquantenhypothese." *Zeitschrift für Physik* 26 (1924a) 178–181.

39 John Stachel. *Einstein from 'B' to 'Z'* 521.

40 Jagadish Mehra and Helmut Rechenberg. *The Historical Development of Quantum Theory, vol. 1, part 2, The Quantum Theory of Planck, Einstein, Bohr and Sommerfeld: Its Foundation and the Rise of its Difficulties 1900–1925* (New York/Heidelberg/Berlin: Springer-Verlag, 1982) 564.

41 S.N. Bose. "Wärmegleichgewicht und Strahungsfeld bei Anwesenheit von Materie." *Zeitschrift für Physik* 27 (1924b) 383–393.

42 William Blanpied. "Satyendranath Bose: Cofounder of Quantum Statistics." *American Journal of Physics* 40 (1972) 1213.

43 See Appendix A.3 for Bose's first letter to Einstein and Appendix A.4 for Einstein's response on 2 July 1924.

44 Satyendranath Bose, "Plancks Gesetz und Lichtquantenhypothese".*Z. Phys.* 26, 178. Hebrew University Doc 7–35.

45 Albert Einstein. "On a Heuristic Viewpoint on the Production and Transformation of Light." *Annalen der Physik* 17, 6 (1905a) 132–148.

46 Albert Einstein. "Uber die Entwickelung unserer Anschauungen über das Wesen und die Konstitution der Strahlung." *Deutsche Physikalische GesellschaftVerhandlungen* 11 (1909b) 482–500.

47 Anthony Duncan and Michel Janssen. "Pascual Jordan's resolution of the conundrum of the wave particle duality of light." *Studies in History and Philosophy of Modern Physics* 39 (2008) 634–666.

48 Ibid. 217; Alexei Kojevnikov. "Einstein's Fluctuation formula and the Wave-Particle Duality." In Yuri Balashov and Vladimir Vizgin (eds.). *Einstein Studies in Russia: Einstein Studies, Vol. 10*. (Boston, Basel: Birkhäuser, 2002) 181–228.

49 For the "shot in the dark" argument see Abraham Pais. *Subtle Is the Lord*: *The Science and Life of Albert Einstein*. (Oxford: Oxford University Press, 1982).

50 Somaditya Banerjee. "Satyen Bose: The Unsung Hero of India." (Paper presented at the Joint Atlantic

Seminar for the History of Physical Sciences (JASHOPS), University of Notre Dame, Indiana, February 4–6, 2005.

51 See Appendix A.3. SNBCS Archive.

52 Ibid.

53 Ibid.

54 See Appendix A.4 for original letter.

55 John Stachel. "Einstein and Bose." In *Einstein from 'B' to 'Z'*. (Boston: Birkhauser, 2002) 519–538.

56 Abraham Pais. "Einstein on Particles, Fields, and the Quantum Theory." In *Some Strangeness in Proportion*. (Boston, MA: Addison Wesley, 1980) 197–251.

57 Ibid.

58 Partha Ghose. "Bose Statistics: A Historical Perspective." In *S.N. Bose: The Man and His Work. Part I: Collected Scientific Papers* (Calcutta: S.N. Bose National Centre For Basic Sciences, 1994) 35–71.

59 Ibid.

60 Ibid.

61 Ibid. 46.
Also see C.V. Raman and S. Bhagvantam. "Experimental Proof of the Spin of the Photon." *Indian Journal of Physics* 6, (1931) 355. Daniel Kennefick an editor of *Collected Papers of Albert Einstein* (CPAE) corroborates my statement through a personal communication in January 2005.

62 Ghose, "Bose Statistics", 46. Reproduced here including typos.

63 C.V. Raman and S. Bhagavantam. "Experimental Proof of the Spin of the Photon." *Nature*, 129 (January 2, 1932) 22–23.

64 Einstein Archive, Cal Tech. Doc. 6-128.

65 Santimay Chatterjee and Enakshi Chatterjee. *Satyendra Nath Bose*. (Calcutta: National Book Trust, 2005) 46.

66 Ibid.

67 Ibid.

68 SNBCS Archives Calcutta Doc 075. Kahn was the brother of Jacqueline Eisenmann, a close friend of Bose in Paris.

69 This fact is based on an undated document in the Jewish National and University Library Jerusalem, of a two-page calculation done by Satyendranath Bose. Scientific Correspondence File Folder 'B-Misc.-II', Call No. 40 1576. As mentioned in *Satyendra Nath Bose: His Life and Times. Selected Works with Commentary*. Ed. Kameshwar Wali. (Singapore: World Scientific, 2009) 313.

70 Ghose. "Bose Statistics" 52.

71 Ibid.

72 Ibid.

73 Ibid. Ghose. "Bose Statistics."

74 SNBCS archives Calcutta Doc 073 (accessed January 2012).

75 John Stachel. "Einstein" 527.

76 Partha Ghose. "S.N. Bose" 57.

77 Ibid.

78 Ghose. "S.N. Bose" 61–62.

79 Ibid.

80 Loudon, Rodney. *The Quantum Theory of Light*. (Oxford: Oxford University Press, 3rd edition, 2000).

81 Partha Ghose (professor of physics and Ph.D. student of Satyendranath Bose), interview by Somaditya Banerjee, January 16, 2012.

82 Ghose (1994) 65.

83 Albert Einstein. "Zur Quantentheorie der Strahlung." *Physikalische Zeitschrift* 18 (1916b) 121–128 (reprint). English translation in Van der Waerden. *Sources of Quantum Mechanics*. (Mineola, NY: Dover, 1968) 63–77.

84 Ghose interview. SNBCS Archives.

85 See Appendix A.6.
86 Ibid.
87 Ghose interview.
88 Just a note of clarification here, that Bose was not heartbroken because of his part-ing in Paris from Jacqueline Eisenmann. He communicated with Eisenmann through letters whenever possible and their friendship lasted well after Bose's first European sojourn. The Bose–Eisenmann relationship was similar to Albert Einstein's with Paul Ehrenfest where both Ehrenfest and Eisenmann were sounding boards for Einstein and Bose respectively.
89 See Appendix A.8.
90 For a similar argument about "electron" see unpublished draft (communicated to author) of Theodore Arabaztis. "Electron's Hesitant Passage to Modernity 1913–1925." And also in Arabatzis. *Representing Electrons: A Biographical Approach to Theoretical Entities*. (Chicago: The University of Chicago Press, 2006).
91 George Basalla. "The Spread of Western Science." *Science* 156 (1967) 611–622.
92 Kapil Raj. *Relocating Modern Science: Circulation and the Construction of Knowledge in South Asia and Europe, 1650–1900*. (London: Palgrave Macmillan, 2007); Homi Bhabha. *The Location of Culture*. (London and New York: Routledge, 1994).
93 Jürgen Habermas. *The Inclusion of the Other: Studies in Political Theory*. (Cambridge: MIT Press, 1998); Robert C. Young. *Colonial Desire: Hybridity in Theory, Culture and Race*. (London: Routledge, 1995).

4 Colonial modernity and C.V. Raman

Verifying the light quantum

In 1930, Nobel Laureate Chandrasekhara Venkata Raman (1888–1970) spoke on the radio for the Indian public, saying:

> I think it will be readily conceded that the pursuit of science derives its motive power from what is essentially a creative urge. The painter, the sculptor, the architect and the poet, each in his own way, derives his inspiration from nature and seeks to represent her through his chosen medium, be it paint, or marble, or stone, or just well-chosen words strung together like pearls on a necklace. The man of science is just a student of nature and equally derives inspiration from her. He builds or paints pictures of her in his mind, through the intangible medium of his thoughts. He seeks to resolve her infinite complexities into a few simple principles or elements of action which he calls the laws of nature. In doing this, the man of science, like the exponents of other forms of art, subjects himself to a rigorous discipline, the rules of which he has laid down for himself and which he calls logic. The pictures of nature which science paints for us have to obey these rules, in other words have to be self-consistent. Intellectual beauty is indeed the highest kind of beauty. Science, in other words, is a fusion of man's aesthetic and intellectual functions devoted to the representation of nature. It is therefore the highest form of creative art.[1]

Raman was a first-generation *bhadralok* scientist, whose experiments at the Indian Association for the Cultivation of Science (IACS) in Calcutta from 1922 onward led to his 1928 groundbreaking discovery of the Raman effect—the frequency-altering scattering of light by atomic systems—for which he was awarded the Nobel Prize in 1930, making him the first "non-Western" scientist to be thus honored.[2] This historic achievement in the sphere of science served as an important political symbol, and a catalyst for Indian strivings toward independence. Though Raman manifested a brand of national consciousness that was different from that of his colleagues Satyendranath Bose and Meghnad Saha, his remark shows his scientific worldview, which integrated concepts of artistic and intellectual beauty. Like the changing patterns on a kaleidoscope, Raman's intellectual interests in science also showed a gradual change, encompassing a broad spectrum.[3] In his early career Raman was interested in acoustics and classical optics, and later in his life after 1930 he showed a fondness for the physics behind crystals.

In the course of his academic career, Raman published more than 480 research papers (both as a single author and as a co-author), many of which appeared in the *Indian Journal of Physics*, which he founded in 1928. He also trained a large number of research students, many of whom went on to hold important portfolios in administration, academia, and politics in their later lives.

Because the reception of Raman's work and early life up to 1928 has already been discussed by Rajinder Singh, some years ago, the present chapter focuses on a social history of how Raman established himself as a key figure in Indian science in the early twentieth century. This chapter examines how Raman sought meaningful connections between a modern scientific worldview and the indigenous knowledge of India by combining his attachment to European science with local intellectual traditions, in order to develop a particular brand of Indian modernity.[4] Specifically, this chapter explores the events that led to the discovery of the Raman Effect by Raman and Kariamanikam Srinivasa Krishnan at the IACS in Calcutta, in February 1928. I shall argue that though the Raman Effect has generally been seen as providing strong evidence for the quantum nature of light, Raman himself was initially a staunch supporter of the classical wave theory. His radical support for the wave theory of light originated from his early career interests in acoustics and Indian musical instruments.

This study will also put Raman's work in the context of the alternate dispersion theories, especially those of Hendrik Antoon Lorentz, Paul Drude, Peter Debye, Arnold Sommerfeld, Charles Galton Darwin, Karl Herzfeld, and Adolf Smekal, as well as scattering experiments by Rudolf Ladenburg and Fritz Reiche, culminating in the dispersion theory of Hendrik A. Kramers.[5]

Raman scattering played an important role in the experimental verification of Kramers' quantum dispersion theory, which formed a conceptual "bridge" between Niels Bohr and Arnold Sommerfeld's "old quantum theory" and Werner Heisenberg's matrix mechanics. The scattering experiments of Russian physicists, Leonid Issakovich Mandelstam and Grigory Landsberg, which were executed around the same time as Raman's experiments in 1928, are also analyzed in this context. Finally, this chapter breaks from the tradition of hagiographic writings[6] on Raman, and argues that Raman had strong networks in the international scientific community, which accounted for his higher popularity in India than that of Satyendranath Bose or Meghnad Saha. Raman's life trajectory also shows the multilayered nature of Indian science, and the intricate nature of the category "(nationalist) science".[7] These layers become especially evident when scholars compare Raman's intellectual style with those of Bose and Saha.

Biographical comments

Born into a middle class *bhadralok* Brahmin family on November 7, 1888, in Tiruchirapalli in the state of Tamil Nadu in South India, Raman was the second amongst eight children. His father, R. Chandrasekaran Aiyar, had accepted the post of lecturer in mathematics and physics at the A.V.N. College in Vizagapatam, when Raman was three years old. Aiyar also excelled in playing Indian musical instruments. Raman's mother, Parvathi Ammal, came from an educated family, known for its reputation in Sanskrit scholarship. At the age of thirteen, Raman went to study at the Presidency College in Madras. After earning the first position in his bachelor's program in 1901, his teachers advised him to travel to England in order to compete for the Indian Civil Service (ICS) examination. When he failed the medical examination, the door to England closed, and feeling relieved, Raman said, "I shall always be grateful to this man [medical officer]."[8] It can be inferred from this remark that either Raman was very much attached to his country, and did not want to serve the British in the ICS, or maybe he had already developed academic interests.

Raman returned to the Presidency College in Madras to work on his master's degree in physics. Attending very few lectures, he devoted most of his time to independent research, focusing mostly on ancient Indian musical instruments. In 1906, he published a short paper in the British *Philosophical Magazine* that analyzed the phenomenon of oblique diffraction using the wave theory of light.[9] Having carefully studied the double-slit diffraction pattern, when light was normally incident at the slits, Raman wondered what would happen when light struck the slits obliquely. He came to the conclusion that when the incident angle was very close to a right angle, the diffraction bands were no longer symmetric, as they were in the case of normal incidence. He then performed simple experiments to verify his conclusions and reported his observations in the *Philosophical Magazine*. As Raman later recalled, he was able to pursue such research, because of the freedom given to him in his education curriculum, particularly because attending lectures was not mandatory. On this matter, Raman said:

> Professor Jones (Professor of Physics) believed in letting those who were capable of looking after themselves to do so, with the result that … I enjoyed a measure of academic freedom that seems almost incredible … During the whole of my two years' work for the M.A. degree, I remember attending only one lecture.[10]

After completing his master's in January 1907, Raman went to Calcutta in eastern India, where he joined the Financial Civil Service as assistant accountant general. Wanting to pursue a research career in physics, Raman pondered over the advantages of being in the administrative service. Such opportunities in administration were open only to the British, and Indians who held British university degrees. But Raman did not have a British university degree. To pursue a research career in the future, and make a living during the intervening period, he had to join the Government service after passing the services entrance exam. Raman said, "I took one look at all the candidates who had assembled there and I knew I was going to stand first."[11] Raman indeed went on to stand first in this examination. His self-confidence, a marked trait of his character, turned out to be well founded in this case. Meanwhile, Raman married a South Indian lady, Lokasundari, a *bhadramahila*,[12] who was later known as Lady Raman.

Raman established contacts with the Indian Association for the Cultivation of Science (IACS), which was founded in 1876 by a noted Bengali *bhadralok* intellectual, Mahendra Lal Sircar, who was a well-known medical practitioner and a philanthropist. Sircar saw scientific expertise and research as important requirements for national awakening. The IACS was the first scientific institution set up in India that aimed to provide opportunities for aspiring Indian scientists in search of active participation in scientific research. Though the Asiatic Society, formed by William Jones in 1784 in Calcutta, was popularly known as the first scientific institute, it was primarily a British society, and natives of India were denied access to it.[13]

As Calcutta offered more job opportunities than other provinces, Raman decided to move to Calcutta in 1907. This move coincided with the rise of the nationalist movement in the city, following the Partition of Bengal by the British in 1905. Calcutta was the capital of British India from 1772 to 1911, when, because of the revolutionary campaigns in the city, the capital was shifted to Delhi in the north. The Partition of Bengal did not have any sustained impact on Raman, and there is nothing in the archives that suggests otherwise. This indifference to the Partition—a major catalyst for Indian nationalism—shows how Raman's identity as an up-and-coming scientist was greater to him than the cause of the nation. This is important to note, as Satyendranath Bose and Meghnad Saha—the other two subjects of this monograph—had reacted very differently to the Partition of Bengal. Because Raman was from South India, which was geographically far away from Bengal, Raman's response was not atypical for someone hailing from a different area of India.

From 1907 to 1917, Raman spent his days in the government office working as an assistant accountant. He devoted days and nights to science. In this period of part-time clerkship and part-time researcher, Raman read Herman Helmholtz's *The Sensations of Tone*, which was translated into English by Alexander Ellis, and published in 1885. Helmholtz's work was presented in a lucid form, specifically for the convenience of music students. It dealt with sound as a sensation and offered many insights that were apparently unclear to Raman, e.g., when Helmholtz admitted that "harmony and quality of tone differ only in degree,"[14] or when he remarked that "the scale best adapted to melody is not adapted to harmony."[15] A reviewer of Helmholtz's article said:

> If, as it appears, the Helmholtzian theories, after twenty-two years of existence and of comment and manipulation by aestheticians, musicians, and physicists, have so far, from a musical point of view, been only destructive in their tendencies and of little direct service to technical theory, it must not be imagined that the assistance of science can be underrated, much less ignored. The beginnings of music are in natural laws; and if we cannot yet say that science follows us in the art to the end, we may say it rejoins us there, and

constitutes the final court of appeal in such ultimate questions, for instance, as the mechanism and genera of scales.[16]

Wanting to explore the ramifications of Helmholtzian wave theories, combined with his interest in the aesthetics of art and science (as we saw in Raman's quotation at the beginning of this chapter), Raman wanted to figure out the acoustics of Indian musical instruments, and check for himself whether the Helmholtzian doctrine of scale, harmony, and melody worked for these instruments. Having difficulty in obtaining access to proper laboratory facilities, Raman chose to make forays in this field of the physics of music. Raman later recalled:

Speaking of the modern world, the supremest figure, in my judgment is that of Hermann von Helmholtz. It was my great good fortune, while I was still a student at college, to have possessed a copy of an English translation of his great work on "The Sensations of Tone." ... It treats the subjects of music and musical instruments not only with profound knowledge and insight, but also with extreme clarity of language and expression. I discovered the book myself and read it with the keenest interest and attention. It can be said without exaggeration that it profoundly influenced my intellectual outlook. For the first time I understood from its perusal what scientific research really meant, and how it could be undertaken. I also gathered from it a variety of problems for research which were later to occupy my attention and keep me busy for many years.[17]

In 1909, Raman was promoted to the rank of currency officer, located to the seemingly "faraway" Rangoon. Frustrated by a lack of scientific equipment, he turned to theory, particularly the wave theory of light. He investigated how the Indian musical instrument *ectara* worked. The *Journal of the Indian Math Club* accepted Raman's theoretical findings on the workings of the *ectara* using oscillations of stretched strings.[18] Raman worked out that for an *ectara*, the period without inhibition of the system sounding-board extensible-wire, for oscillations parallel to the length of the wire, is half that of the period of transverse oscillation of the wire.

In a similar fashion, Raman studied other musical instruments—the violin, sitar, tambura, and the veena—while he also analyzed their frequency response and found the dependence of the emission of various frequencies on the bowing pressure, the normal modes of vibration, and various harmonics, using the procedure of calculations as shown earlier. Raman's early fascination with acoustics became the basis for his later insights into the nature of light. His attachment to the wave theory stemmed from his initial interests in the physics behind several Indian musical instruments like the *ectara*. Regarding music, stringed instruments, and culture in ancient India Raman stated:

Music, both vocal and instrumental, undoubtedly played an important part in the cultural life of ancient India. Sanskrit literature, both secular and religious, makes numerous references to instruments of various kinds, and it is, I believe, generally held by archaeologists that some of the earliest mentions of such instruments to be found anywhere are those contained in the ancient Sanskrit works. Certain it is that at a very early period in the history of the country, the Hindus were acquainted with the use of stringed instruments excited by plucking or bowing, with the transverse form of the flute, with wind and reed instruments of different types and with percussion instruments.[19]

Speaking about percussion instruments as a wave theorist, Raman noted:

As is well known, the vibrations of a circular stretched membrane or drum-head excited by impact are generally of an extremely complex character. Besides the gravest or fundamental tone of the membrane, we have a large retinue of overtones which stand to each other in no sort of musical relation. These overtones are always excited in greater or less degree and produce a discordant effect. All the instruments of percussion are known to European physicists in which a circular drum-head is employed have therefore to be regarded more as noise producers introduced for marking the rhythm than as musical instruments. This is true even of the kettle-drum which is tuned to a definite pitch and occasionally used in European orchestral music. All the instruments of percussion known to European science are thus essentially non-musical and can only be tolerated in open air music or in large orchestras where a little noise more or less makes no difference. Indian musical instruments of percussion however stand in an entirely different category. Times without number we have heard the best singers or performers on the flute or violin accompanied by the well-known indigenous musical drums, and the effect with a good instrument is always excellent. In was this, in fact, that conveyed to me the hint that the Indian instruments of percussion possess interesting acoustic properties, and stimulated the research.[20]

Here, Raman speaks about the subtleties of Indian music, especially the percussion instruments, contextualizing them with respect to European orchestral music. He claimed that these nuances of Indian music inspired him to delve deeper into these subjects. Raman published thirty scientific papers during this period in such journals as the *Journal of the Indian Math Club*, *Nature*, *Philosophical Magazine*, and *Physical Review*.[21] As a result, he was offered the Palit Professorship of Physics at the Calcutta University in 1917 by Ashutosh Mukherjee, the vice-chancellor at that time. Though Raman's new job came with considerably lower pay than the accountant's job, he accepted it. Now he could devote more time to teaching and research at Calcutta University, and to experimental work at the IACS. Ramaseshan, a student of Raman, described his teacher's life in those days in the following passage:

5.30 a.m. Raman goes to the Association. Returns at 9.45 a.m., bathes, gulps his food in haste and leaves for office, invariably by taxi [horse-drawn carriage] so that he might not be late. At 5 p.m., Raman goes directly to the Association [IACS] on the way back from work. Home at 9.30 or 10 p.m. Sundays, whole day at the Association.[22]

During that period, Raman also developed an odd habit of wearing a headband. In South India, people normally did not wear such headbands. Headbands, or turbans, as they are popularly called in India, are worn by people from the northern part, especially from the state of Punjab and parts of Rajasthan. While speaking at a conference in 2006, M.S. Swaminathan, one of Raman's contemporaries, recalled Raman's ready wit when someone asked him why he wore a turban. Raman replied, "Oh, if I did not wear one, my head will swell. You all praise me so much and I need a turban to contain my ego."[23] Hence, this turban story is yet another indication of Raman's eagerness to be different. From this story, one can conclude that, for Raman, the turban symbolized Indianness or a distinctiveness that made him look different from his colleagues, both Indian and non-Indians (see Figure 4.1).

En route towards the discovery of the Raman Effect

From "isolated" Rangoon, Raman was glad to receive a transfer to Calcutta where he joined the up-and-coming IACS in 1911. Raman's travels to Rangoon and back to Calcutta included sea voyages, during which he spent considerable time pondering over the sea and its colors. At the IACS, Raman wanted to diversify his research portfolio. Making a transition from wave acoustics to optics made sense. The diversity of Raman's interests in optics ranged from the visualizations of the sea to astronomical optics. For example, he studied Saturn, and gave two lectures on his observations of the interference fringes and diffraction patterns of two light sources using the wave theory of light. In 1912, Raman helped mount a telescope on a small wooden observatory on the roof of IACS. Having had some background in studying Saturn, he then turned the focus of his research to Jupiter's surface saying, "I think the problem of scattering of light by a planetary body is not altogether an easy one and there may be room for further investigations here."[25]

Thus, Raman's initial interests in acoustics, and his research findings regarding the workings of *ectara* and other Indian percussion instruments by using the wave theory, served as the background for his later interests in light scattering at the IACS. G.N. Ramachandran, writing in the journal *Current Science*, remarks:

> The study of acoustics is intimately connected with the study of vibrations and waves, and it is not surprising that Raman's interests passed from his early love for acoustics on to a life-long devotion to optics, the other great domain of classical wave mechanics. In fact, if one may talk of a unifying trend in the scientific work of Raman, it may be said to reside in the study of wave phenomena.[26]

Figure 4.1 Raman wearing his turban.[24]

While Raman worked in Calcutta at the IACS, he did not have the support of an entourage of assistants. There was only one assistant, Ashutosh Dey (another *bhadralok*), who helped him set up and carry out experiments (see Figure 4.2). In the wake of the Partition of Bengal, the field of education showed an upward trend with native Indians taking recourse to all possible means for national upliftment. The distinguished educator Ashutosh Mukherjee played an important role in this crucial period. He was appointed as the vice-chancellor of Calcutta University in 1906. In 1908, he set up the Calcutta Mathematical Society, which was to be a forum for research and teaching in mathematics and physics.

Mukherjee's efforts led to philanthropists like Taraknath Palit, Rashbehari Ghosh, Maharaja of Darbhanga, and the Maharaja of Khaira to generously donate

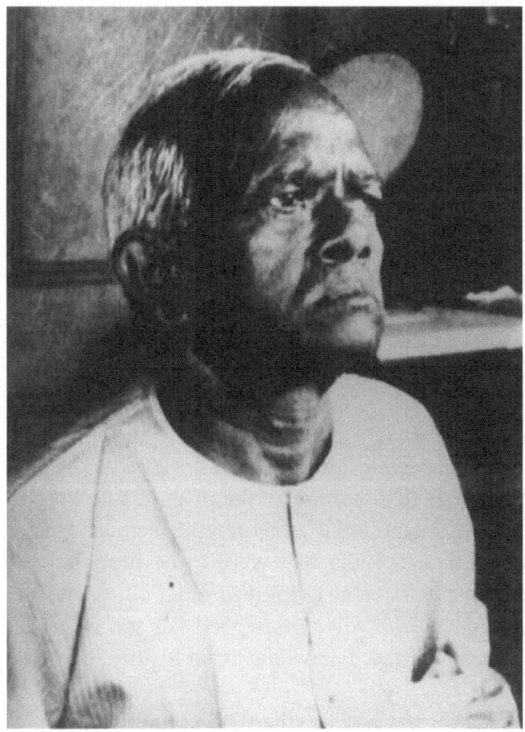

Figure 4.2 Raman's lone assistant at IACS: Ashutosh Dey.[27]

funds for opening the University College of Science (UCS), and for subsequent endowment-chairs to be held by Indian scientists. Raman made a name for himself in acoustics as well as astronomical optics, and became a stalwart in the institutional milieu of the IACS. Despite his achievements, Raman was not necessarily the best choice for Mukherjee, because of the presence of Jagadish Chandra Bose (JCB), who had already established himself as a celebrated scientist and a physics professor at the Presidency College in Calcutta. However, it is unclear why JCB was not granted the professorship position over Raman. Though Raman was already in the financial civil service, which paid a higher salary, he accepted the position offered to him, since this position at the UCS also entailed teaching—a profession he had longed for.[28] Unfortunately, he became involved in a conflict with scientists from Bengal like JCB, who wrote to the Vice-Chancellor of Calcutta University:

> It has been reported to me that, on the 25th instant, a member of the Department of Physics of the University College of Science (UCS) called at my Laboratory at the Presidency College during my absence, and with special instructions from Prof. Raman to invite my senior mechanic to transfer

his services to the College of Science Physical Department, with offer of increased salary above what he gets from me ... even up to three times if necessary ... I must, therefore, formally express to the University my regret.[29]

These grievances against Raman were part of a larger problem in the history of Indian science, that of regionalism.

On March 26, 1914, Raman received a letter from the registrar of Calcutta University that said:

> I am directed to inform you that the Hon'ble the Vice Chancellor and Syndicate agree to the condition on which you are prepared to accept the appointment of Sir Taraknath Palit Professor of Physics, namely, that during your incumbency you will not be required to leave India to proceed to any foreign country.[30]

Raman responded in the affirmative, accepted the offer, and resigned from his government position. But due to the logistical paraphernalia of the Palit endowment, he could not join immediately. The colonial government intervened, and was reluctant to fund endowment-chairs in India that were to be occupied by native Indians. However, by 1917, Raman became employed as a Palit Professor of physics. As an occupant of that post, with a well-equipped lab and research grants for building instruments, Raman started a new chapter in his life in optics and light scattering.

Raman also received access to the labs at IACS, where he used to work on a part-time basis during his tenure as a financial clerk. Meanwhile, a research group began to grow around Raman in Calcutta. As Raman earned nationwide fame for his research and teaching prowess in Calcutta at UCS and IACS, several students joined his group from South India (University of Madras), which included his key collaborator, Kariamanikam Srinivasa Krishnan, and also K.R. Ramanathan, L.A. Ramdas, K.S. Rao, Sunderaraman, V.S. Tamma, Y. Venkataramayya, A. Ananthakrishnan, S. Bhagavantam, A.S. Ganesan, C. Ramaswamy, S.S.M. Rao, S. Paramasivan, N.S. Nagendra Nath, C.S. Venkateswaran, and S. Venkateswaran, who eventually became Raman's research assistants.[31]

Interestingly, most of Raman's assistants were South Indians, and it can be inferred that Raman had a preference for choosing assistants from South India, his native land. Raman's feelings for South Indians may be viewed as an index of subnationalism, which espoused his support for the interests of the people of South India. Subnationalism and regionalism in the case of Raman were very closely related, as we will see later.

While at the IACS, Raman had occasional contentions with Meghnad Saha, and the eminent mathematician, D.N. Mallik. The conflict between Raman and Saha originated in 1917, when the former had held the first Palit Professor of Physics position, a coveted position in the Calcutta University. This was a time when Raman attempted to limit the membership of IACS to only people from Southern India, creating problems for the Institute and other senior members like

Jagadish Chandra Bose, Kedareswar Banerjee, Panchanon Das, and Manindra Nath Mitra who were not from the South.[32] Saha was the leader of this group of Raman's opponents. The conflict between Saha and Raman reflected the tension between regional interests, South Indian versus Bengali identity and autonomy.

Saha expressed his annoyance on several occasions regarding the favoritism exercised by Raman toward people from South India at the expense of qualified non-South Indians. Saha was also apprehensive of the fact that Raman could jeopardize the future prospects of Saha's students. When advising Pratap K. Kichlu, an upcoming scientist from North India, Saha said, "When you submit [your] thesis for [the] D.Sc … the examiners ought to be Professor [Ralph H.] Fowler, Lord Rayleigh and myself. Do not allow Raman or [John W.] Nicholson to be put in."[33]

Apart from that unpleasant social clash with fellow scientists in the sphere of academics, Raman came into conflict with D.N. Mallik over the interpretation of Fermat's Law. Mallik published a mathematical paper on theoretical optics in the *Bulletin of the Calcutta Mathematical Society* in July of 1913 on the kinetic nature of optical energy. Mallik concluded in his paper that "we must have for light propagation, $T - V = $ Constant,"[34] where T is the kinetic energy and V the potential energy so that Hamilton's Principle and Fermat's Law might be consistent with each other. Raman objected to this statement and remarked, "this statement of Dr. Mallik is most seriously in error, maybe shown in a very simple and general manner."[35]

Raman showed the fallacy of Mallik's contention and concluded by saying:

> I would suggest that Dr. Mallik should read Huyghens' own statement of the case in his original treatises on Light. Dr. Mallik might possibly also obtain some clearer ideas on the relation between Hamilton's Principle and Optical theory by reading Wangerin's exposition of the work of Voigt on the subject. [Encyclopaedie Der Math Wiss., Band V, Art. 39.] From the authoritative reference quoted above, it will be seen that Dr. Mallik is in error when he assumes (without any analytical justification) that $T - V = $ constant for an optical medium. Such an assumption is wholly unnecessary and leads to results which are quite meaningless.[36]

This episode indicates Raman's grasp in theoretical optics in his early days as a scientist, as he was quite well read in classical optics, especially the works of Christiaan Huygens, the Dutch scientist of the seventeenth century. At a personal level, this conflict shows Raman's commanding nature, and quite dismissive tone towards the senior Indian (especially Bengali) colleagues like Mallik. While showing mastery over theoretical topics in classical optics, Raman wasted no time in planning a research program in physics with experimental skill.

Having acquired a good number of assistants, Raman started consolidating his research program in Calcutta by building instruments and probing the subtleties of wave optics for the purpose of understanding the molecular basis of the macroscopic phenomenon of refraction. In 1919, he began developing an interest in the molecular diffraction of light. With B.B. Ray, Raman published a paper on a light-scattering problem, where a beam of light was sent through a solution in which sulphur suspension particles were formed. Here, the two scientists observed a counterintuitive phenomenon.

The intensity of the transmitted light decreased as the solution became gradually turbid, which was quite intuitive, but with the gradual passage of time there was a gradual reappearance of transmitted light by the suspension.[37] Raman tried to explain this apparently strange phenomenon with the help of Fresnel and Huygens wave theory by arguing that the reappearance of transmitted light occurs when the growth in size of the suspension particles leads to forward scattering and interference in a forward direction. These events were the background for Raman's later researches into light scattering.

In 1921, Raman received an opportunity to visit England for the first time and attend the University Congress at Oxford as a representative of Calcutta University. When Raman was transferred to Rangoon earlier in his life, his travel methods included a sea voyage, during which he pondered over the optics of the sea, based on his research experiences in music and acoustics. On his considerably longer return from England on board the *S.S. Narkunda*, Raman further contemplated the beautiful blue color of the sea.[38]

As he was initially interested in issues surrounding the beauty, the aesthetics, and the connections between art and science, explaining the color of the sea was a natural outgrowth of Raman's pedagogical interests. Earlier, in 1899, Lord Rayleigh had successfully explained the blue color of the sky by giving a scattering formula for a gas,[39] and had explained the color of the sea simply by arguing that the sea was blue because it reflected the color of the sky. Rayleigh scattering involved scattered radiation with the same frequency as the incident radiation. Such scattering, in which the frequency does not change, is called *coherent*. In this context, Rayleigh gave the following remark in the *Royal Institution Proceedings*:

A recent voyage round Africa recalled my attention to interesting problems connected with the colour of the sea. They are not always easy of solution in consequence of the circumstance that there are several possible sources of colour whose action would be much in the same direction. We must bear in mind that the absorption, or proper, colour of water cannot manifest itself unless the light traverse a sufficient thickness before reaching the eye. In the ocean the depth is of course adequate to develop the colour, but if the water is clear there is often nothing to send the light back to the observer. Under these circumstances the proper colour cannot be seen. The much admired dark blue of the deep sea has nothing to do with the colour of water, but is simply the blue of the sky seen by reflection. When the heavens are overcast the water looks grey and leaden; and even when the clouding is partial, the sea appears grey under the clouds, though elsewhere it may show colour. It is remarkable that a fact so easy of observation is unknown to many even of those who have written from a scientific point of view.[40]

As Raman had experienced sea voyage, Rayleigh's explanation of the color of the sea unsettled Raman. Considering these misgivings, Raman stated:

> Observations made in this way in the deeper waters of the Mediterranean and Red seas showed that the color, so far from being impoverished by suppression of sky-reflection, was wonderfully improved thereby … It was abundantly clear from the observations that the blue color of the deep seas is a distinct phenomenon in itself, and not merely an effect due to reflected skylight. When the surface-reflections are suppressed the hue of the water is of such fullness and saturation that the bluest sky in comparison with it seems a dull grey...The question is: What is it that diffracts the light and makes its passage visible? An interesting possibility that should be considered is that the diffracting particles may, at least in part, be the molecules of the water themselves.[41]

Raman's reasoning also relied on the Einstein–Smoluchowski formula of 1910 that explained critical opalescence, which is the strong scattering of light by a medium near a phase transition. Einstein's key insight in this paper was that the phenomena of critical opalescence and the blue color of the sky, though not related to each other, were both due to density fluctuations caused by molecular constitution of matter.[42] What happened to light scattering when the medium was not close to a phase transition? In the case of light scattering from solids, what frameworks have to be taken? These and similar problems attracted Raman and his associates. As a result, Raman published in *Nature* in 1922 on the color of the sea.[43] The following results were found:

Raman concluded that

> It is evident from these figures that the blue of the sea would be much more saturated than the blue of the sky, which is the standard of comparison. The height of the homogenous atmosphere being 8 kilometres, the sea would be about half as bright as the zenith sky on a clear day. This agrees well with the photometric determinations made by Luckiesh during aeroplane flights over deep ocean water in the Atlantic (*Astrophys. J.*, 49, 1919, p. 129). Luckiesh makes it clear that the greater part of the observed luminosity of water viewed perpendicularly really arises from light diffused upwards from within the water. His determinations thus appear to furnish a quantitative proof of the theory which attributes the colour of the deep sea to molecular scattering of light.[44]

While exploring the subtleties of light scattering in liquids, Raman's framework espoused few distinct effects that cause and enhance the scattering, namely the density fluctuations in a fluid, and also the non-spherical nature of the molecules constituting the fluid. On performing experiments at the IACS, with his collaborators, Ramanathan, Krishnan, Ramdas, Ganesan, Seshagiri Rao, Venkateswaran,

Kameswara Rao, Ramakrishsna Rao, and Ramachandra Rao, Raman found that scattering from transparent liquids always contained some radiation of frequency lower than that of the incident light. Such a scattering, in which the frequency does change, is called non-coherent. Important in this context are the observations of Ramanathan in 1923.

Building an international image

Though Raman had a peculiar fondness for his own country, his patronage network in science led him to think beyond the nation. As soon as he had a well-equipped laboratory with logistical support and a research group, he started planning international trips. Even while starting out as Palit Professor at UCS, he visited London to attend the science Congress. In 1924, he was in a better position as a scientist than in 1921, when he was still building his reputation. He could travel undisturbed, while his research groups back home worked on the problems he directed them to address. In 1924, Raman received an invitation to attend a meeting of the British Association for the Advancement of Science (BAAS) in Canada. His travels took him to North America for the first time.

In August 1924, Raman was in Toronto, giving talks on his research at IACS on light scattering. After his Canadian sojourn, he went to the Franklin Institute in Philadelphia for its centenary celebrations. Robert Millikan invited him to visit Caltech, where he stayed for three months. Here, Raman spoke with astrophysicist Svein Rosseland about getting recognized for his researches. Raman's immediate scientific goal was to "make a great discovery and receive the Nobel Prize."[45] Departing from California toward India, he visited Sweden, Denmark, and Germany, and returned to Calcutta in March 1925.[46]

The 1920s were a very fertile period for the development of physics on a transnational scale. The Compton Effect, which was observed by Arthur Holly Compton in late 1922, established that a quantum of radiation underwent a discrete change in wavelength when it experienced a billiard-ball collision with an electron at rest in an atom. Compton's X-ray scattering experiments confirmed this discrete change in wavelength ($\lambda^1 - \lambda$).[47] Contrary to classical electromagnetism, where the wavelength of scattered rays is equal to the initial incident wavelength, Compton's experiments showed a wavelength shift given by the equation.

$$\lambda^1 - \lambda = \frac{h}{mc}\left(1 - \cos\Theta\right)$$

This phenomenon soon became known as the Compton Effect, and Compton received the Nobel Prize (1927) for his discovery. This phenomenon provided the experimental proof for quanta, and convinced most physicists of the reality of light quanta. However, the results were not universally accepted because of a few skeptical critics. For instance, the Harvard physicist, William Duane, expressed doubts about Compton's results at the British Association for the Advancement of Science (BAAS) meeting in Toronto in 1924, where Raman was also present

(Figure 4.3).[48] The reporter for *Nature* went into more detail on the background and character of the Toronto debate, saying:

> Duane found that, with his apparatus, he was unable to find evidence for the existence of the effects observed by Compton. Compton, on the other hand, could not repeat satisfactorily Duane's experiments. Prof. Raman made an eloquent appeal against a too hasty abandonment of the classical theory of scattering … The fundamental difference between the two theories remain; Duane uses only the well-established quantum energy equation, while Compton in addition introduces the idea of conservation of momentum in the interaction between radiator [sic] and matter.[50]

Raman further took Compton to task at the meeting and said, "Compton, you're a very good debater, but the truth isn't in you."[51] This statement can be taken as evidence that Raman was unmoved by the Compton Effect and continued to believe in waves.[52] Though he tried to downplay the Compton Effect and its conceptual significance, Compton's insights at the Toronto debate were very much present in Raman's framework in light scattering. He conceptualized his researches as an optical analogue of the Compton Effect, and remarked:

Figure 4.3 Raman with Compton at the center.[49]

Its real significance as a twin brother to the Compton effect first became clear to me at the end of 1927 when I was preoccupied with the theory of the subject. I regarded the ejection of the electron in the Compton effect essentially as a fluctuation of the atom of the same kind as would be induced by heating the atom to a sufficiently high temperature, and the so-called directed Compton effect as merely an unsymmetrical emission of radiation from the atom which occurs at the same time as the fluctuation in its electrical state. The conception of fluctuation is a very familiar one in optical and kinetic theory, and in fact all our experimental results in the field of light scattering has been interpreted with its aid. There was, therefore, every reason to expect that radiations of altered wavelength corresponding to fluctuations in the state of scattering molecules should be observed also in the case of ordinary light.[53]

Clearly Raman was influenced by Compton's experiments, as the latter remarked "that it was probably the Toronto debate that led him to discover the Raman Effect two years later."[54]

In 1927, Ramanathan observed that when sunlight passed through a scattering medium, a small fraction of light scattered with a change of frequency. Though very feeble, it could still be observed. All of Raman's collaborators agreed that the mechanism producing the modified radiation was fluorescence, which was primarily produced as a result of impurities in the liquids that act as scattering centers. Scientists attempted to purify the material by distillation, yet the modified non-coherent radiation persisted. This radiation was termed by Ramanathan as "feeble fluorescence," for a lack of a more appropriate name.[55] The only quandary was that this modified radiation was polarized, while the fluorescent radiation is not polarized. When Raman pondered on these observations, he realized that what Ramanathan called "feeble fluorescence" was not fluorescence at all, but a form of modified scattering. Raman recalled:

At a very early stage of our investigations, we came across a new and entirely unexpected phenomenon. As early as 1923, it was noticed that when sunlight filtered through a violet glass passes through certain liquids and solids, e.g., water or ice, the scattered rays emerging from the track of the incident beam through the substance contained certain rays not present in the incident beam. The observations were made with colour filters. A green glass filter was used which cut off all light if placed between the violet filter and the substance. On transferring the glass to a place between the substance and the observer's eye, the track continued to be visible though feebly. This is a clear proof of a real transformation of light from a violet into a green ray. The most careful chemical purification of the substance failed to eliminate the phenomenon.[56]

An obstacle in the way of a speedy progress of scattering experiments in Raman's lab at IACS was the feebleness of radiation. The introduction of any optical element like a prism or a phase retarder would have made the signal weaker. Techniques were needed to improve the intensity of the incident radiation. A primary light source for Raman was sunlight, along with a suitable combination of filters, to differentially select parts of the spectrum to be analyzed. A monochromatic light source was important to analyze the wavelength-shifted scattering. Some of the logistical impediments were cleared when Raman acquired a seven-inch refracting telescope, which, in tandem with a lens having a small focal length, could condense a beam of sunlight into a high intensity pencil of radiation. Equipped with the required optical apparatus, Raman and his associates could analyze the "feeble fluorescence" in the beginning of 1928.[57]

These conceptual detours formed the basis of Raman's interests in light scattering from a liquid, which he pursued from 1919 to 1927, and which culminated in the celebrated Raman Effect. Krishnan observed the same effect in scattered light of sixty-five different purified liquids leading to Raman's observation in glasses in late 1927.

Explaining the effect

The theoretical explanation of the Raman Effect followed the discovery of its phenomenology. According to the current understanding, the Raman Effect occurs when a single frequency (monochromatic) beam of light (or a light quantum) strikes a scattering medium like benzene, and, in the process, collides with the molecules of the liquid by either giving up some energy or collecting some energy from it. When the phenomenon, as seen through the scattered radiation coming out of the sample, is analyzed, one observes *coherent* as well as *non-coherent* radiation. The coherent radiation is the Rayleigh scattered terms, while the non-coherent radiation consists of some modified frequencies, i.e., either a lower frequency (Stokes terms) or a higher frequency (anti-Stokes terms). This phenomenology, which physically manifests itself by a change in frequency, becomes the observable quantity for the experimenter.

While Figure 4.4 shows the phenomenon of Raman scattering, Figure 4.5 and Figure 4.6 illustrate modern-day explanations of the Raman Effect in terms of light quanta. Figure 4.5 is an energy level diagrammatic explanation of the Raman Effect. The ground state and the excited states are shown as bands between which transitions are made.

The simplest possible transition is the Rayleigh one that arises because of the polarizability of the molecule. This process involves an excitation from the ground state to the excited state, and a subsequent de-excitation back to the original ground state. This phenomenon physically manifests itself by scattered radiation of the same frequency as the incident one. It is the changes in polarizability (electric dipole moment) during molecular motions that are responsible for the Stokes and anti-Stokes line, and therefore the Raman Effect. The Stokes transition can be explained by saying that there is an excitation at a particular frequency

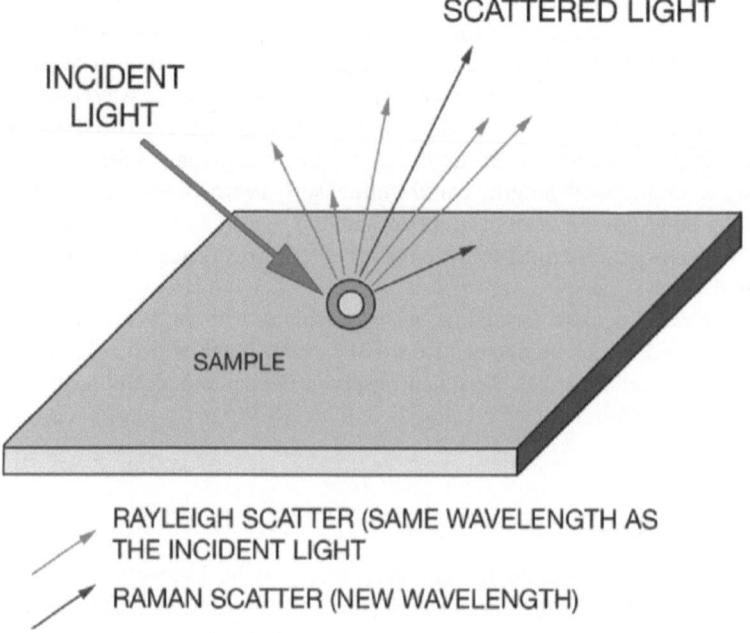

Figure 4.4 Raman scattering.[58]

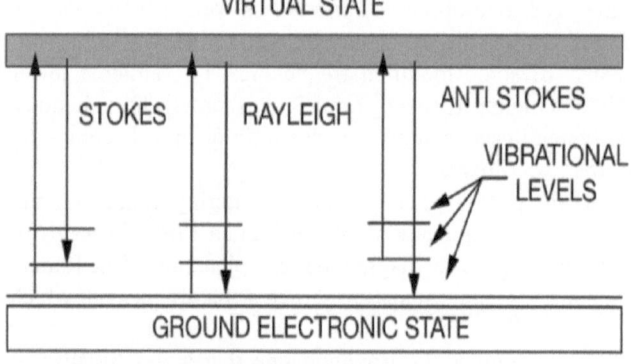

Figure 4.5 Energy level diagram showing Rayleigh and Raman (Stokes and anti-Stokes) scattering.[59]

from the ground state, and a subsequent de-excitation to a state of lower frequency (increased wavelength) than the initial one. This implies that the scattered photon has a lower energy than that of the incident photon. This Stokes scattering was first proposed theoretically by Austrian physicist, Adolf Smekal, in 1923 in *Die Naturwissenschaften.*[61]

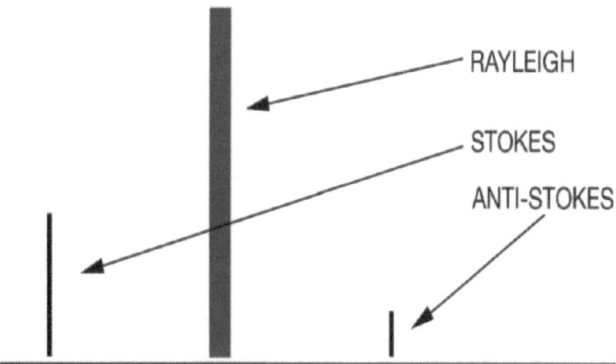

Figure 4.6 Comparison of Rayleigh with Raman spectrum with its Stokes and anti-Stokes lines.[60]

It may be pointed out here that Smekal was a firm believer in Einstein's light quantum, and he suggested a corpuscular theory of dispersion. In 1923, Smekal showed that scattered monochromatic light would consist of coherent terms as well as non-coherent terms. While the anti-Stokes radiation can be explained by noting that the exciting transition is already from an excited state, the subsequent de-excitation is at a higher frequency, and, moreover, a higher energy by the relation E= hv. However, as the transition starts out in a state where sufficient vibrationally excited molecules might not be present, the anti-Stokes line is therefore weaker than the Stokes line, as also seen in Figure 4.6. The previous figure (see Figure 4.5) also illustrates that the Stokes and anti-Stokes line is equally displaced from the Rayleigh line. This happens because in both cases one vibrational quantum of energy is gained or lost. The Raman Effect occurs when a photon is excited on a molecule and interacts with the polarizability of the molecule. Classically, it can be viewed as a perturbation of the molecule's electric field. The spectral shifts of the modified radiations give one a measure of the rotational or vibrational frequencies of the molecule.

The newspaper clipping in Figure 4.7 was the first media announcement of the discovery of the Raman Effect that took place on February 28, 1928.[62] The newspaper announcement also spoke about the Compton Effect, which was considered a radical breakthrough for light quanta. Raman was influenced by Compton's work as we have seen earlier, but he tried to downplay the "revolutionary" aspect of the work, especially in the verification of light-quantum at the Toronto debate. In fact, when Krishnan informed Raman in 1927 that Compton had been awarded the Nobel Prize, Raman remarked, "If this is true of X-rays, it must be true of light too … We must pursue it and we are on the right lines. It must and shall be found. The Nobel Prize must be won."[63] As we will see in the next section, the meanings of light quanta were quite different in India. Through Raman's interpretive lens, one can find a manifest ambiguity in the explanation of the Raman Effect.

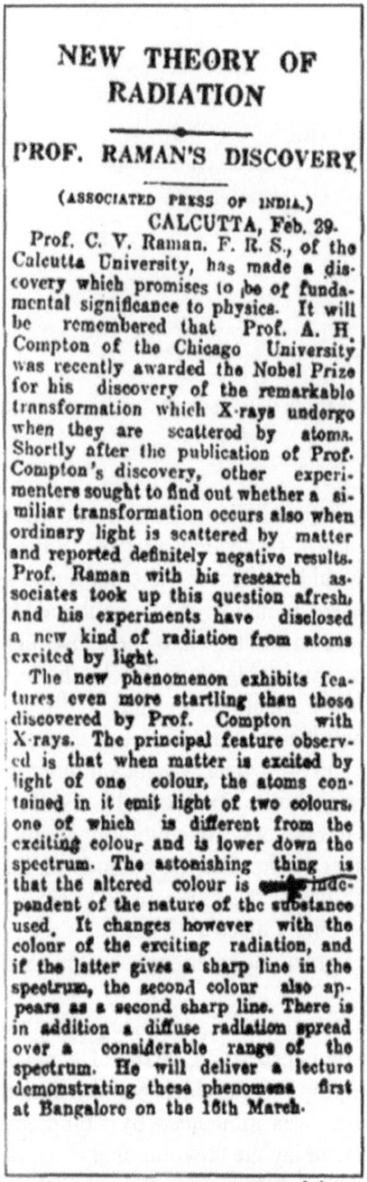

NEW THEORY OF RADIATION

PROF. RAMAN'S DISCOVERY

(ASSOCIATED PRESS OF INDIA.)

CALCUTTA, Feb. 29.

Prof. C. V. Raman. F. R. S., of the Calcutta University, has made a discovery which promises to be of fundamental significance to physics. It will be remembered that Prof. A. H. Compton of the Chicago University was recently awarded the Nobel Prize for his discovery of the remarkable transformation which X-rays undergo when they are scattered by atoms. Shortly after the publication of Prof. Compton's discovery, other experimenters sought to find out whether a similiar transformation occurs also when ordinary light is scattered by matter and reported definitely negative results. Prof. Raman with his research associates took up this question afresh, and his experiments have disclosed a new kind of radiation from atoms excited by light.

The new phenomenon exhibits features even more startling than those discovered by Prof. Compton with X-rays. The principal feature observed is that when matter is excited by light of one colour, the atoms contained in it emit light of two colours, one of which is different from the exciting colour and is lower down the spectrum. The astonishing thing is that the altered colour is quite independent of the nature of the substance used. It changes however with the colour of the exciting radiation, and if the latter gives a sharp line in the spectrum, the second colour also appears as a second sharp line. There is in addition a diffuse radiation spread over a considerable range of the spectrum. He will deliver a lecture demonstrating these phenomena first at Bangalore on the 16th March.

*First newspaper announcement of the
Discovery of the Raman Effect
made on 28th Feb. 1928*

Figure 4.7 First newspaper announcement of the discovery of the Raman Effect made on February 28, 1928.[64]

Raman Effect and Quantum Physics

How did Raman account for the effect? Raman offered the following explanation in February 1928:

> If we assume that the X-ray scattering of the unmodified type observed by Prof. Compton corresponds to the average state of the atoms and molecules, while the 'modified' scattering ... corresponding to their fluctuations from that state, it would follow that we should expect also in the case of ordinary light two types of scattering, one determined by the normal optical properties of the atoms and molecules, and another representing the effect of their fluctuations from their normal state ... The subject of light scattering is thus a meeting ground for thermodynamics, molecular physics and the *wave-theory of radiation*.[65]

Here, Raman is talking in terms of the wave theory of radiation, reflecting a bias which may have come from his early association with ancient Indian musical instruments, as has been argued at the outset. Raman also re-derived the Compton shift in 1928 with the classical theory, explaining it through the Doppler Effect, which can be taken as compelling evidence for Raman's faith in the wave theory. There is, however, some ambivalence in his understanding of this novel effect, as revealed in his remark on March 16, 1928, at a lecture in Bangalore:

> As a tentative explanation, we may adopt the language of the quantum theory, and say that the incident quantum of radiation is partially absorbed by the molecule, and the unabsorbed part is scattered. The suggestion does not seem to be altogether absurd and indeed such a possibility is already contemplated in the *Kramers-Heisenberg theory of dispersion*.[66]

The "quantum of radiation" mentioned in the above quotation is just a quantity of energy in the form of classical radiation. Bohr uses the term like that in his 1913 paper. Despite appearances to the contrary, the remark quoted above really does not mean that Raman subscribed to the notion of the light quanta. There is some disagreement about this meaning in the historiography. Rajinder Singh says that "well before Raman discovered the Raman Effect, he accepted the quantum nature of light."[67] However, Abha Sur claims that "Raman himself was a quintessential classical physicist certainly in his training and even more so in his outlook."[68]

Nonetheless, this debate in historiography asks the bigger question about the Raman Effect's connections to the experimental verification of the revolutionary formalisms of quantum mechanics that were unfolding in the mid-1920s in Europe. As Thomas Kuhn later remarked about the Kramers–Heisenberg paper and their treatment of the Smekal–Raman incoherent scattering terms: "you get what you would now recognize as cross-products terms in a matrix expansion and that is what inspired matrix mechanics."[69]

The place of the Raman Effect in the history of quantum physics

Meanwhile, physicists in Europe grappled with the Rayleigh-like coherent terms in the scattered radiation in old quantum theory.[70] In the classical Lorentz–Drude picture of dispersion, an electromagnetic wave of frequency v strikes a one-dimensional simple harmonic oscillator with characteristic frequency "v0.". What happens next depends on whether or not "v" is close to "v0." The Lorentz–Drude dispersion formula has resonance poles at the frequency "v0." As long as "v" is far removed from "v0," one is in the regime of so-called *normal* dispersion; close to "v0," one is in the regime of *anomalous* dispersion.

In 1915, Peter Debye and Arnold Sommerfeld proposed a dispersion formula similar to the classical Lorentz–Drude formula in the context of Bohr's new quantum model of the atom. The resonance poles in the Sommerfeld–Debye formula are at the orbital frequencies in the Bohr atom. This could not be reconciled with the experimental data, which clearly showed that the poles should be at the radiation frequencies, which, in the Bohr model, differ sharply from the orbital frequencies.

In the early 1920s, several alternative dispersion theories were proposed that addressed this problem. In 1922, using light-quanta, Charles Galton Darwin introduced a damping and interference mechanism, and argued that though light from a single atom would have the orbital frequency, the interference of an ensemble of waves led to scattered light waves having the radiation frequency.[71] However, conservation of energy only held statistically in his model. Furthermore, Bohr pointed out that Darwin's theory failed when considering low intensity light.

Meanwhile, Karl Herzfeld suggested a mechanism for obtaining non-coherent scattered radiation.[72] Using light-quanta, Herzfeld argued that the stationary states allowed by the quantum conditions were not the only permissible ones. There were orbits of all sizes and shapes corresponding to all values of the constants of integration, which resulted in a "diffuse- quantization" with indeterminate energy values. This was a variant of the work by Bohr and Sommerfeld, and their quantization condition. Hence, the orbits not obeying the quantum conditions were assumed to have a very small a-priori probability, and electrons could remain in them for about a femtosecond.[73]

In 1923, Adolf Smekal described a new type of quantum transition, which he called "translational quantum transitions," that one obtained from scattering monochromatic radiation from atoms.[74] Smekal wrote, "Because of the change in direction of the radiation effected by them [i.e., by the translational quantum transitions], we shall speak in the case m = n about normal dispersion and in the case m ≠ n about anomalous dispersion."[75] Note that Smekal used the terms "normal dispersion" and "anomalous dispersion" in an idiosyncratic way and that the distinction he made is usually labeled coherent versus non-coherent. Smekal's view was opposed to that held by Niels Bohr, who was a stubborn supporter of the wave nature of radiation. This view became important for the later development of dispersion theory by Kramers and Heisenberg in 1925, and later in 1928 when Raman and his associates would discover important data in their light-scattering experiments. It is, however, unknown when (if at all) Raman became aware of Smekal's work, and how he responded to it. It can be inferred that Raman's complete faith in wave theories and natural distrust of the light quantum could have made him ignore Smekal's work in *Naturwissenschaften*.

Post-factum, Smekal's paper was often quoted in the literature as indicating a prediction of the Raman Effect. For example, Austrian scientist K.W.F Kohlrausch published a book entitled *Der Smekal-Raman-Effekt* in 1931. Ramdas, one of Raman's students at IACS, commented in 1928 that Smekal's paper did not appear to have been noticed by any experimental physicist working in the field of light scattering, including the group working under Raman. But Ramdas also noted that Kramers and Heisenberg took notice of Smekal's idea, and further developed it in their treatment of the quantum theory of scattering in 1925.[76]

In this context, it is interesting to explore the work of Kramers and Heisenberg, as they were, like Raman, using only wave theory of light, and the experiments on dispersion by Rudolf Ladenburg and Fritz Reiche at Breslau.[77] Regarding these experiments, Schrödinger remarked:

> The existence of this remarkable kind of secondary radiation … has not yet been demonstrated experimentally. The present theory also shows distinctly that the occurrence of this scattered radiation is dependent on special conditions, which demand researches expressly arranged for that purpose … For the extraordinary scattered radiation, which is to be discussed, is proportional to the product of the spontaneous emission coefficients in question.[78]

The main object of Kramers and Heisenberg's paper was to account for the non-coherent scattering suggested by Smekal, without taking recourse to light quanta and using only the wave theory.[79] The Kramers–Heisenberg paper was also the first systematic exposition of the new theory for coherent scattering that Kramers had presented in two short notes to *Nature* in 1924.

The theory of dispersion by Kramers and Heisenberg replaced the unsatisfactory Sommerfeld–Debye theory. The key ingredients of Kramers' dispersion theory were Einstein's A and B coefficients, and Bohr's correspondence principle. Kramers built on the work that the Breslau experimentalist, Rudolf Ladenburg, had done in 1921, with important help from the Breslau theoretician Fritz Reiche.[80] Ladenburg used a dispersion formula with poles at the observed radiation frequencies. Anthony Duncan and Michel Janssen argue that Ladenburg's "main contribution was when he recognized that the oscillator strengths corresponding to various transitions could all be interpreted in terms of transition probabilities, given by Einstein's A and B coefficients."[81] Ladenburg's formula of classical oscillator strengths and quantum transition probabilities A's are the Einstein coefficients, P is the polarization, N the number of atoms. Ladenburg's formula was valid only for the ground state.[82]

$$P_r(t) = \frac{N_r c^3 E}{32\pi^4} \sum_s \frac{A_{s \to r}}{v_{s \to r}^2 \left(v_{s \to r}^2 - v^2\right)} \cos 2\pi v t \qquad (4.1)$$

For the excited state, one needed two terms and that is what Kramers derived, basing his derivation on Ladenburg's insights from the above equation. Also, for using the correspondence principle, which holds only for highly excited states, one needed two states. This specification gives the following equation:

$$P_r(t) = \frac{N_r c^3 E}{32\pi^4} \left(\sum_{s>r} \frac{A_{s\to r}}{v_{s\to r}^2 \left(v_{s\to r}^2 - v^2 \right)} - \sum_{t<r} \frac{A_{r\to t}}{v_{r\to t}^2 \left(v_{r\to t}^2 - v^2 \right)} \right) \cos 2\pi v t \quad (4.2)$$

Orbits do not correspond to observable quantities but transitions do; for example, the frequency terms between transitions from "s" to "r" and intensity are observable through the corresponding Einstein coefficients "A."

Ladenburg replaced the numbers of oscillators in the classical Lorentz–Drude formula by transition probabilities in the Bohr atom, given by Einstein's emission and absorption coefficients. Ladenburg's extensive experiments on dispersion in gases since 1908 had convinced him that the resonance poles of the dispersion formula had to be at the radiation frequencies, even though he and Reiche saw no way of deriving this result from quantum theory.

In 1924, Kramers finally accomplished this task on the basis of Bohr's correspondence principle. Kramers found that the formula suggested by Ladenburg needed to be supplemented by a second term, which would only contribute appreciably to the dispersion if a substantial fraction of the atoms were in an excited state. In the late 1920s, Ladenburg and his collaborators tried to experimentally verify this second term in the Kramers dispersion formula.

The reason why Raman Effect is important in this dispersion work by Ladenburg and Kramers is because of the second term of the Kramers dispersion formula (equation 4.2). Though Ladenburg and Reiche tried to verify the second term experimentally, they did not succeed. It was the Raman Effect that provided the experimental confirmation of the second term of Kramers' dispersion formula. Princeton physicist Francis Low explains that:

> Raman found that light scattered by certain substances may have a slightly changed color from the original light beam. This effect is hard to account for according to nineteenth century physics, whereas it may be definitely predicted on the basis of the new quantum theory, of which it is therefore an important experimental confirmation.[83]

In essence, Raman did associate his findings of light scattering with the Kramers dispersion formula. Krishnan's personal diary, where he kept notes of daily scientific events, revealed an exchange of views between Raman and his associates before the discovery of the Raman Effect. Specifically, the following entry from 1928 shows this exchange of views:

Feburary 7, 1928

After meals at night, Venkateswaran and myself were chatting together in our room when Prof. (Raman) suddenly came to the house (about 9 pm) and called for me. When we went down, we found he was much excited and had come to tell me that we had observed that morning must be the Kramers-Heisenberg effect we had been looking for all these days. We therefore agreed to call the effect *modified scattering*. We were talking in front of our house for more than a quarter of an hour when he repeatedly emphasized the exciting nature of our discovery.[84]

Thus, it is evident that Raman was aware of the work of Kramers and Heisenberg. But there is no evidence that Raman was aware of Smekal's theoretical insights in the early 1920s. Rajinder Singh, however, has argued both ways. In an earlier paper, he argued that Raman used Kramers' theory to interpret the experimental results. But later, Singh argued that Raman was unaware of the work of Kramers and Heisenberg, and said, "none of this theoretical work (of Kramers and Heisenberg) ... exerted a direct influence on the discovery of the Raman effect."[85] This apparent uncertainty of whether or not Raman was aware of earlier theoretical work feeds into bigger questions of originality and recognition in the history of science. While Raman might very well have been aware of Kramers' earlier work, as suggested in Krishnan's diary, it could be inferred that Raman tried to build an image to the contrary, perhaps in pursuit of the Nobel Prize.[86]

Work by Landsberg and Mandelstam: The simultaneous discovery of Raman Effect, the Nobel Prize of 1930, and Stigler's Law of Eponymy[87]

Often physicists and historians regard the Nobel Prize as an index of a research program's success and modernity. It has been recently argued that, as opposed to the physics of principles (espoused by Einstein, Planck, and Bohr), the physics of problems as practiced by the Sommerfeld school could make a strong claim to have been the most successful research program for theoretical physics in the twentieth century, because at least eight Nobel laureates were associated with it.[88]

The Nobel Prize is commonly seen as the final authority for assessing the success or failure of a research program. This is, however, a highly reductionist view. According to Robert Friedman, this stereotype overlooks the politics and the hidden agendas associated with the prize. Friedman shows, in his *The Politics of Excellence*, how simplistic such stereotypical claims are regarding the Nobel Prize: "Without understanding the limitations and weaknesses of the process, the recipients were afforded instant prestige as part of the Nobel cult."[89] Behind the Nobel Prize given to Raman were factors that corroborate Friedman's argument in this case.

Around the same time in 1928, when a novel scattering mechanism, subsequently known as the Raman Effect, was discovered in Calcutta on February 28, this very mechanism was also discovered in Moscow on February 21. A group of Russian physicists had been working on similar scattering experiments as Raman. Grigory Samuilovich Landsberg and Leonid Isaakovich Mandelstam attempted to elucidate the fine structure of the Rayleigh line induced by modulation of scattered light with Debye thermal waves.[90] Unlike Raman, Landsberg and Mandelstam used quartz as their scattering medium. Quartz was not as easy to find as benzene or the other aromatic compounds that were the scattering medium of Raman.

The work by R.J. Strutt,[91] who studied light scattering in quartz and concluded that what he had observed was not light scattered from quartz molecules, but light reflected from false scattering centers, was the basic motivation for Landsberg. The latter took up this task of studying molecular light scattering in real crystal, and proposed a criterion for the differentiation of scattered light and reflected light from false scattering centers.

Meanwhile, Mandelstam theoretically calculated the change in the light frequency as given by:[92]

$$\Delta\Omega = \pm 2n\omega \frac{V}{c} \sin\frac{\theta}{2}$$

Here n, ω, V, c, θ are the refractive index, angular frequency, velocity of sound, speed of light and scattering angle. Landsberg and Mandelstam published their results in *Naturwissenschaften* on July 13, 1928. They said:

> In the investigation of molecular scattering of light in solids which we undertook to find out whether a change in wavelength occurs that might be expected in the framework of the Debye theory of heat capacity, we ran into a new phenomenon which seems to us to be of certain interest. The phenomenon consists in a change of wavelength whose value however has an order of magnitude and origin other than we had expected.[93]

Landsberg and Mandelstam argued that the non-Rayleigh modified scattering terms, as seen by the satellite lines, was due to the interaction between the light and infrared molecular vibrations. Fabelinskii, who was a student of Mandelstam, reports that the first observations of his mentors were on February 21, 1928, which was a week before those of Raman and his collaborators. Landsberg and Mandelstam, however, published their work on July 13, 1928, a few months after their discovery. Apparently, the main reason for the delay was that Gurevich, a relative of Mandelstam, was arrested and sentenced to death. As a consequence, Mandelstam had to take a break from research and spend more time to mitigate the death sentence. In the end, however, he succeeded in reversing the verdict of the death sentence. Gurevich was exiled to the city of Vyatka. His life was saved, but at the expense of the publication of Landsberg and Mandelstam's innovative work.[94]

Mandelstam wrote to physicist Orest Khvolson, saying, "We first noted the appearance of the new lines on February 21, 1928. On a negative from an experiment of February 23–24 (exposure time 15 hours) the new lines were clearly visible."[95] Fabelinskii argues that Landsberg and Mandelstam reported their discovery at the beginning of August 1928, at the sixth Congress of the Association of Russian Physicists. Twenty-one of the 400 participants at the Congress were foreign scientists, including Max Born, Brillouin, Darwin, Debye, Dirac, Phol, Pringsheim, Philip Frank, and Scheel. Darwin wrote, "Perhaps the most interesting work is that of Prof. Mandelstam and Landsberg. The latter described how they had independently discovered Raman phenomenon, the scattering of light with changed frequency."[96] Max Born remarked:

> The effect discovered by Landsberg and Mandelstam in crystals is essentially identical to the effect observed by Raman and his colleague Krishnan in liquids. Russian physics can justly take pride in the fact that this important discovery was made by the Moscow researchers independently of the Indians and nearly simultaneously (February 20, 1928). This coincidence is one more demonstration of the international nature of our science, which now spans the entire world.[97]

In fact, Raman's students, A. Jayaraman and A.V. Ramdas, wrote on Raman's centenary about this simultaneous discovery saying, "Really the Raman Effect was independently discovered by Landsberg and Mandelstam in calcite and quartz crystals."[98] Though Mandelstam and Landsberg saw the novel scattering phenomenon a week before Raman, the Nobel Prize in Physics in 1930 went to the latter. One may wonder about the reasons behind such an incident.

There were twenty-one nominations for the Nobel Prize in 1930, and Raman was proposed ten times, either as a single candidate or jointly with his collaborators.[99] Figure 4.8 shows the people who had nominated Raman for the Nobel Prize in 1930. It can be inferred that, as Raman established contacts with scientists in Germany, England, France, Sweden, and North America, he was better known internationally than Mandelstam and Landsberg. M. Siegbahn and C.W. Oseen, both members of the Nobel physics committee in 1930, knew Raman personally.

Proposed by	The Candidate/s
E Bloch (Paris)	WR Wood* & CV Raman
N Bohr (Copenhagen)	Wood or Wood & Raman
O Chwolson (Leningrad)	Half for Raman and the rest for Landsberg & Mandelstam
J Perrin (Paris)	Raman or Raman & Heisenberg
FL de Broglie (Paris)	Raman
HM de Broglie (Paris)	Raman
R Pfeiffer (Breslau)	Raman
J Stark (Grosshesselohe)	Raman
E Rutherford (Cambridge)	Raman
CTR Wilson (Cambridge)	Raman

Figure 4.8 Nominations for the 1930 Nobel Prize in physics.[100]

An interesting exchange of letters in 1928–29 between Raman and Niels Bohr summarizes the story. In a letter to Bohr in 1928, Raman remarked:

> The great kindness you have shown me in the past encourages me to make a request of a personal character. As you know, my work on the new radiation effect has been received with enthusiasm in scientific circles, and I feel sure that if you give your influential support, the Nobel Committee for physics may recommend that the award for 1930 may go to India for the first time. The proposal for the award has to reach the Nobel Committee before 31 January 1930. I have greatly hesitated in writing to you about this, and it is only because I felt sure that you sympathise with the scientific aspirations of India that I have ventured to do so.[101]

As a matter of fact, Bohr was influenced by Raman's letter, and he extended his support for Raman through his nomination. This support played a key role in Raman winning the prize. The physics-prize verdict of 1930 also proves a variant of statistician Stephen Stigler's Law of Eponymy, that "no scientific discovery is named after its discoverer." Though this idea is quite a generalization, in Raman's context it can be said that many simultaneous scientific discoveries are not named after all their discoverers (Figure 4.9).[102]

Figure 4.9 Raman (second from right) with Niels Bohr to Raman's left. The others from the left are George Gamow, Thomas Lauritsen, T.B. Rasmussen and Oskar Klein.[103]

Arnold Sommerfeld's visit to India in 1928 coincided with Raman's explorations in light scattering.[104] Sommerfeld repeated Raman's experiments at the IACS, and verified these experiments. Through Sommerfeld, his colleagues in Munich and Berlin came to be aware of Raman's work. Strangely enough, Raman did not even mention the names of Mandelstam and Landsberg in his Nobel acceptance speech in 1930. A part of this speech is given below:

> The general principle of correspondence between the quantum and classical theories enunciated by Niels Bohr enables us, on the other hand, to obtain a real insight into the actual phenomena. The classical theory of light scattering tells us that if a molecule scatters light while it is moving, rotating or vibrating, the scattered radiations may include certain frequencies, different from those of the incident waves. This classical picture, in many respects, is surprisingly like what we actually observe in the experiments. It explains why the frequency shifts observed fall into three classes, translational, rotational and vibrational, of different orders of magnitude. It explains the observed selection rules, as for instance, why the frequencies of vibration deduced from scattered light include only the fundamentals and not the overtones and combinations which are so conspicuous in emission and absorption spectra. The classical theory can even go further and give us a rough indication of the intensity and polarization of the radiations of altered frequency. Nevertheless, the classical picture has to be modified in essential respects to give even a qualitative description of the phenomena, and we have, therefore, to invoke the aid of quantum principles. The work of Kramers and Heisenberg, and the newer developments in quantum mechanics which have their root in Bohr's correspondence principle seem to offer a promising way of approach towards an understanding of the experimental results.[105]

Although there is no mention of the Russian physicists who had observed the phenomenon before Raman, there are two references to Bohr, who had played an important role by nominating Raman for the Nobel Prize as earlier mentioned. The quotation also shows Raman's fondness for classical wave theories, of which Bohr was a radical supporter.

As argued earlier, Raman's proximity to Indian musical instruments, and attachment to the works of German polymath Helmholtz, were some of the reasons for his fondness for classical wave theories. If Raman had eventually accepted the light quantum, it would have been a hesitant acceptance with the disclaimer that classical theories were more fundamental. And, in the case of large quantum numbers, according to Bohr's correspondence principle, quantum calculations had to agree with classical calculations. Moreover, one could be well in favor of quantum theory, and still be against the concept of light quantum, something that was common to both Bohr and Raman. Though the new quantum mechanics of the mid-1920s were mostly a German phenomenon, its leading exponents, such as Arnold Sommerfeld, were keenly interested in Raman's works in light scattering.

Sommerfeld and the reception of Raman's work in Germany: Orientalism and science

Sommerfeld was a great admirer and supporter of Indian physicists and their works.[106] He was attracted to J.C. Bose's work in electrophysiology, Saha's work on stellar spectra, Satyendranath Bose's work on quantum statistics, and Raman's work on light scattering. The *Zeitschrift für Physik* was the channel through which Sommerfeld gained familiarity with the work of Indian physicists. Sommerfeld asked Saha to give a lecture in Munich in 1921, and Saha obliged. Raman, along with Saha, invited Sommerfeld to visit India and give lectures at the University of Calcutta. Sommerfeld visited India in 1928 after the discovery of the Raman Effect, and gave talks mostly on atomic structure and wave mechanics in Calcutta. While in India, Sommerfeld wrote an article praising modern Indian science, and equated its quality with that of Europe and America. Sommerfeld expressed special admiration for Raman's discovery, and for Saha's work in astrophysics.[107]

The Raman Effect, however, did not get a good reception within certain sections of the German physics community. Göttingen physicist Otto Blumenthal, Georg Joos at the University of Jena, along with Richard Gans were all apprehensive of Raman's work. Gans in particular had a negative view about Indian scientists, apparent in his writing to Sommerfeld from Jena on May 14, 1928, when he says,

> Do you think that Raman's work on the optical Compton effect in liquids is reliable? To repeat the experiment is not a big task and most probably we are going to do it. The sharpness of the scattered lines in liquids seems doubtful to me.[108]

Goos based his ideas on an unsuccessful repetition of the Raman Effect at the University of Munich. As Singh noted, "Gans had a negative opinion about Indian scientists ... and had a skeptical attitude towards the quality of publications by Indian physicists ... and also told Sommerfeld that Indian physicists are not reliable."[109]

On June 9, 1928, Sommerfeld wrote to Joos that "in my opinion Raman is correct and important. He writes to me, that the difference between the lines is exactly equal to the infra-red frequencies of the molecules under consideration."[110] Thus, Sommerfeld's response to Indian science provides an alternate perspective that reconstructed the socio-scientific image of India as not exclusively spiritual but also scientific. Following Raman, one can infer that Indian science did not follow the Western trajectory to modernity, but an alternate path that encompassed ideas about the human spirit, the virtues of human endeavor and achievement, and a search for truth for its own sake. Raman himself thought:

> In my case strangely enough it was not the love of science, nor the love of Nature, but an abstract idealization, the belief in the value of the Human Spirit and the virtue of Human Endeavor and Achievement. When I read

Edwin Arnold's classic *The Light of Asia*, I was moved by the story of the Buddha's great renunciation, of his search for truth, and of his final enlightenment. It showed me that the capacity for renunciation in the pursuit of exalted aims is the very essence of human greatness.[111]

This line of thought is striking, because Raman was moved by a Western account of Oriental wisdom, thereby revealing the contradictory nature of his personality. He seemed to have developed an aversion for the British, and yet was fond of other Europeans like Sommerfeld and Arnold (British though he was). Raman's quotation and his scientific work also call into question certain stereotypes of an opposition between Oriental and Western thought.

If, as Singh asserts, Gans was prejudiced against Indian scientists, the controversy among German physicists about Raman's work may have involved their various preconceptions about Oriental science.[112] In my view, the defining characteristic of Raman was that, even though he was a major harbinger of modernity in Indian society, he tended to reject the Oriental stereotypes in the West that would separate and oppose modern science to traditional Oriental knowledge. Upon Raman's return to Calcutta after receiving the Nobel Prize, Lady Raman said that her husband had "sought to dispel the notions in Europe that India was rather too 'Spiritual.'"[113] Raman's interest in Indian classical musical instruments evidences his fascination towards Indian tradition. Yet his light-scattering experiments advanced the most modern European science.

Raman vacillated between tradition and modernity, but his characteristic approach was a combination of the two. Before his discovery of the Raman effect in 1928, he re-derived the Compton scattering wavelength using wave theory. Raman's attitudes regarding the traditional and the modern were ambivalent, even contradictory. His apparently strange outlook espoused a methodology that broke away from negative stereotypes of Oriental science, and, instead, adopted a variant of what Richard G. Fox has called "affirmative orientalism."[114] By this phrase, Fox suggests that "Orientalist narratives were appropriated by Indian intellectuals and applied in such a way as to undercut the colonialist agenda."[115] Hence, such narratives did not operate in straightforward and orderly fashions, but illustrated some of the ambiguities of colonial physics in early-twentieth-century India.

Raman's extensive institutional, personal, and pedagogical networks were similar to those of Western scientists, even though he developed them while working in a colonized, non-Western nation. Then too, in contrast to Orientalistic assumptions of Eastern inferiority, several Western scientists, such as Sommerfeld, helped reconfigure myths about the East by highlighting the scientific achievements of Raman and other scientists of his generation, who were working in the Orient. Sommerfeld convinced his colleagues in Germany of the authenticity of Raman's works, especially after his visit to India. Sommerfeld's India visit paved the way for several collaborations between physicists at the University of Calcutta, and Sommerfeld's Munich school. Ramesh Chandra Majumdar, a graduate student in Calcutta University, was awarded the Zeiss scholarship by the Deutsche Akademie to do research in Munich. Several Indian students from Calcutta studied at the University of Munich under the guidance of Sommerfeld, Walther Gerlach, Thierfelder, and Schmauss, the noted meteorologist. Sommerfeld received the honorary D.Sc. degree from the University of Calcutta in 1928.[116]

Raman himself visited Munich as a Nobel Laureate in 1930. In 1934, when he became the director of the Indian Institute of Science (IISc), Sommerfeld recommended one of his students named Ludwig Hopf, who happened to be a Jewish refugee, to teach at the IISc. Raman's endeavor was instrumental in the creation of a special readership in theoretical physics at IISc from October 1935 to March 1936. This readership went to the Jewish scientist Max Born, who sought refuge after his dismissal from the University of Göttingen.[117]

Between nationalism and regionalism

Robert Anderson has argued that as the national scientific community was developing during the 1930s, communications amongst Indian scientists within different regions increased considerably. Researchers interacted with each other more frequently on a regional and national basis; travel by train was more frequent, and slightly easier for scientists; the postal and telegraphic system continued to improve; and opportunities for both status and power arose that were not just local in character.[118]

It is debatable whether Raman was a nationalist, but his personality had a peculiar brand of sensitivity for his nation that can be seen from his exchanges with some of the institutions and colleagues in the West. Speaking at the convocation address to the students of Benaras Hindu University in 1926, Raman spoke about his speeches while he was in Europe and said:

> Do you think I spoke about Madras or of Calcutta? No! I spoke of Kashi, of Benaras, of the historic city on a ridge overlooking the Ganges which stands at the very heart of India, as the living centre of our ancient culture and learning. I spoke of the new University that has sprung up, so fitly, at this age-old seat of learning and is the living embodiment of the aspirations of new India … It is not the function of a university to grow bookworms. The function of a university is to train men to serve their country and above all to train those who can become leaders, leaders of science, leaders of industry, leaders in all other fields of activity … A bookworm consumes books but produces only dust. A true scholar does not merely consume knowledge but also produces knowledge.[119]

On May 15, 1924, Raman was elected as Fellow of the Royal Society of London. Kameshwar Wali argued on the basis of his conversations with another Indian Nobel laureate, S. Chandrasekhar, concerning Raman's unhappiness over an article published in the London *Times*, circa 1967, on the Nobel Laureate Fellows of the Royal Society, because it did not mention Raman's name.

Raman blamed the omission on the Society and wrote to P.M.S. Blackett, who was the president of the Society at that time, saying that unless he were given a satisfactory explanation for this omission, he would resign; which he did in March 1968 after Blackett's response.[120] Rajinder Singh, however, argues that there was no communication between Blackett and Raman, and there was also no such list of Fellows of the Royal Society who won a Nobel Prize published in the London *Times* between 1967 and 1968. Singh concludes that "Raman's resignation remains a mystery."[121]

Though this is an apparently strange episode, Raman had developed an awareness about his nation, and a national identity that was not atypical of scientists in late colonial India. In an undated quote on his feelings on receiving the Nobel Prize, Raman remarked:

> When the Nobel award was announced I saw it as a personal triumph, an achievement for me and my collaborators – a recognition for a very remarkable discovery, for reaching the goal I had pursued for seven years. But when I sat in that crowded hall and I saw the sea of faces surrounding me, and I, the only Indian, in my turban and closed coat, it dawned on me that I was really representing my people and my country. I felt truly humble when I received the Prize from King Gustav; it was a moment of great emotion but I could restrain myself. Then I turned round and saw the British Union Jack under which I had been sitting and it was then that I realized that my poor country, India, did not even have a flag of her own – and it was this that triggered off my complete breakdown.[122]

However, by examining Raman's character closely, one can conclude that Raman's nationalist inclinations in colonial India might have been a reason behind this feeling. Therefore, Raman's resignation can also be viewed as a protest against a seemingly "discriminatory" act on the part of the British. There is evidence, however, that Raman used to be a difficult person to get along with as well as quite arrogant, which added a peculiar dimension to his character. Fabelinskiy describes a particular incident and explains that:

> in 1957 Raman visited Moscow to receive the Lenin Peace Prize. He was invited to read a lecture about his theory of solids at a seminar run by P.L. Kapitza at the Institute of Physical Problems. I attended the seminar. Some 15–20 minutes into the lecture, L.D. Landau, sitting in the front row made a remark. Raman appeared to have nothing to say in response. Instead, he began shouting, stamping his feet, swinging his arms, insulting Landau and talking rot. Landau stood up and left the conference hall. The chairman did not utter a word. I have never seen the like of that.[123]

Furthermore, when C.G. Darwin expressed skepticism during a visit to Raman's laboratories in 1935, Raman said "it is far easier to straighten the tail of a dog than to try to convince an Englishman of the correctness of [one's] theories."[124] As Raman pursued modern science in a colonial environment under the British Raj, it's possible that he developed a feeling of cynicism and a lack of fondness toward the English in particular.

Despite such occasional disagreements and seemingly quarrelsome behavior in Raman's life, one should not be hasty to categorize him as "*abhadra,*" or ungentlemanly. Raman's achievements in his early days as a scientist at the IACS, where he successfully built a group of early-career scholars leading to his Nobel winning work, and his later move to the Indian Institute of Science and the Raman Research Institute in Bangalore, trumps any other anomalous behavior he might have had.

Raman showed a fondness for his nation that is harder to classify as "nationalist" compared to the sentiments of Satyendranath Bose and Saha.[125] His nationalistic sentiments were expressed through his emotions while accepting the Nobel Prize in 1930, and his later resignation from the Royal Society. His symbolic gestures, like wearing indigenous headgear, projected an attitude that was nationalist, but not staunchly anticolonial.[126]

Interestingly, Raman's worldview resonated with those of the German, Helmholtz, the Briton, Rayleigh, and the Dane, Bohr. Raman combined European science, such as the classical wave theories of Huygens, Fresnel, Helmholtz, and Rayleigh, with local intellectual traditions of Indian music, fusing them into a specific brand of Indian modernity that emerged in the case of the Raman Effect. His early fascination with acoustics became the basis of his later insights into the nature of light, as well as his ardent support for the wave theory of light, and his ambivalent outlook toward the quantum.

Raman's career trajectory also shows the multilayered and multidimensional nature of Indian science. Not all Indian scientists thought alike, and there were occasional disagreements between Raman and J.C. Bose, Saha, and Mallik, and even with Western scientists like Born and Compton. I consider these differences as regionalism (on a local and global scale) – the regional prioritizing of traditions, personal networks, and solidarities. In spite of employing plenty of opportunities available for scientific research and teaching at the Calcutta University and the IACS, Raman never identified himself as a scientist from Bengal. Most of his associates were from South India, so when he was offered a position at the IISc in Bangalore in 1931, he was quick to take it and leave his established position in Calcutta.

This chapter also locates the Raman Effect in the history of quantum mechanics by putting his work on the dispersion of light in the context of the alternative dispersion theories of Lorentz–Drude, Debye–Sommerfeld, C.G. Darwin, Herzfeld, Smekal, and the scattering experiments by Ladenburg and Reiche which culminated in the dispersion theory of Kramers and Heisenberg. Raman scattering played an important role in the verification of quantum mechanics by confirming experimentally the second term of the Kramers–Heisenberg dispersion formula.

Scientific image-building was also a matter of concern for Raman. For this purpose, he made educational pilgrimages to Europe and North America where he developed a dialogue with his Western colleagues, such as Compton, Millikan, Rosseland, Bohr, and Sommerfeld. These apparently scientific internationalist gestures helped Raman win the Nobel Prize in 1930, even though the Russian physicists Mandelstam and Landsberg had observed the novel scattering mechanism before Raman.

Finally, Raman's worldview reconfigured Orientalist stereotypes by presenting his interest in science as a pursuit of truth for aesthetic and intellectual satisfaction and a seemingly Weberian idea of "science as a vocation."[127] More generally, through the lens of a social history of Raman's life, one can conclude that science in India did not follow the Western trajectory to modernity. Instead, it opened up an alternative path that encompassed ideas about modernity in conjunction with Indian tradition.

Notes

1 Chandrasekhara Venkata Raman. *New Physics: Talks on Aspects of Science.* (Freeport, NY: Books for Libraries Press, 1951) 135–142.
2 http://www.nobelprize.org/nobel_prizes/physics/laureates/1930/, accessed on January 10, 2012.
3 Somaditya Banerjee. "C.V. Raman and Colonial Physics: Acoustics and the Quantum." *Physics in Perspective* 16, 2 (2014) 146–178.
4 Rajinder Singh. "C.V. Raman and the Discovery of the Raman Effect." *Physics in Perspective* 4, 4 (2002) 399–420.
5 Peter Debye. "Die Konstitution des Wasserstoff-molekuls." Sitzungsberichte der mathematisch- physikalischen Klasse der Kniglichen Bayerischen Akademie der Wissenschaften zu Munchen. (1915) 1–26; Paul Drude, *Lehrbuch der Optik.* (Leipzig: S. Hirzel, 1900), English transl.: *The Theory of Optics.* transl.: C. R. Mann and R. A. Millikan. (New York: Longmans, Green, 1902); Arnold Sommerfeld. "Die Drudesche Dispersionstheorie vom Standpunkte des Bohrschen Modelles und die Konstitution von H2, O2, and N2." *Annalen der Physik* 53 (1917) 497–550; K.F. Herzfeld, "Versuch einer quantenhaften Deutung der Dispersion", *Zeitschrift fur Physik* 23 (1924) 341–360; A. Smekal, "Zur Quantentheorie der Dispersion". *Die Naturwissenschaften* 11 (1923) 873–875; R. Ladenburg, "Die quantentheoretische Dispersionsformel und ihre experimentelle Prufung". *Die Naturwissenschaften* 14: (1926) 1208–1213; F. Reiche, and W. Thomas. "Uber die Zahl der Dispersionselektronen, die einem station ären Zustand zugeordnet sind." *Zeitschrift fur Physik* 34 (1925) 510–525; H.A. Kramers and W. Heisenberg. "Uber die Streuung von Strahlung durch Atome." *Zeitschrift fur Physik* 31 (1925) 681–707. Page references to English translation in Van der Waerden, 1968, 223–252. B.L. van der Waerden. *Sources of Quantum Mechanics.* (Amsterdam: North Holland Pub. Co, 1967).
6 Singh. "C.V. Raman," 399–420; G. Venkataraman. *Raman and his Effect.* (Hyderabad: Universities Press, 1995); Uma Parameswaran. *C.V. Raman: A Biography.* (New Delhi: Penguin Books, 2011).
7 Pratik Chakrabarty. *Western Science in Modern India: Metropolitan Methods, Colonial Practices.* (New Delhi: Permanent Black, 2004) 180–210.
8 Venkataraman. *Raman and his Effect* 3.
9 Raman. "Unsymmetrical Diffraction Bands Due to a Rectangular Aperture." *Phil. Mag.* 1906 (6) 12, 494–498.
10 Venkataraman. *Raman* 5.
11 Ibid. 6–10.
12 Female analogue of a *bhadralok.*
13 IACS Archives (accessed June 2012). Sircar established the Calcutta Journal of Medicine in 1868 and was an influential popularizer of Indian science. Also see Gyan Prakash. *Another Reason* 59.

14 Alexander J. Ellis (trans.). *On the Sensations of Tone as a Physiological Basis for the Theory of Music.* (London: Longman, Greens, 1885) 481–484.
15 Ibid.
16 Ibid.
17 C.V. Raman. *Books That Have Influenced Me*: *A Symposium* (Madras: G. A. Natesan & Co.,1947) 21–29.
18 C.V. Raman, "The Ectara." *J. Indian Math. Club* (1909) 170–175.
19 C.V. Raman. *Sir Ashutosh Mookherji Silver Jubilee Volume* Vol. 2. (Calcutta: Calcutta University Press, 1922) 179.
20 Ibid. 180–185.
21 http://www.vigyanprasar.gov.in/scientists/cvraman/raman1.htm (accessed December 5, 2011).
22 Venkataraman. *Raman* 6.
23 See internet resource www.thehindu.com/2006/06/21/stories/2006062107600200.ht m (accessed on (March 5, 2007). This being the online edition of one of India's national newspaper *The Hindu.*
24 Raman Research Institute Digital Depository Archives http://dspace.rri.res.in/bitst ream/2289/5667/11/CVR273br.jpg (accessed January 15, 2019).
25 *Report of Astronomical Society*, April 1913; Parameswaran. *Raman* 66.
26 G.N. Ramachandran. *Current Science* 40 (1971) 212.
27 Proceedings of the *IACS*, Vol 2, 1917. Courtesy IACS Archives.
28 See Singh. *Raman* 399–420.
29 J.C. Bose to D.P. Sarbadhikari, August 30, 1917 (private copy) as quoted in Singh. *Raman* 399–420.
30 Parameswaran. *Raman* 80. Just to clarify here that technically Raman could travel outside India which he did in 1921 when he went to England.
31 Ibid. 94.
32 IACS archives. Also in http://hdl.handle.net/10821/285 (accessed on January 6, 2012).
33 M.N. Saha to P.K. Kichlu, August 15, 1927, Nehru Archives (Saha papers), New Delhi (accessed June 2012).
34 Calcutta Mathematical Society Archives, Kolkata, Doc B.1913.
35 Ibid.
36 Calcutta Mathematical Society Archives, Doc. B. 1917.
37 C.V. Raman and B.B. Ray, *Proceedings of the Royal Society of London* A100 (1921) 102–109. The strange reappearance of color was as follows: being at first indigo, then blue, blue-green, greenish-yellow, and finally white.
38 Venkataraman. *Raman* 34.
39 The scattering coefficient was inversely proportional to the fourth power of wave-length, see for example Rodney Loudon (2000) 374.
40 Lord Rayleigh. *Royal Institution Proceedings. Nature* LXXXIII (February 25, 1910) 48   http://archive.org/stream/scientificpapers05rayliala/scientificpapers05rayliala_ djvu.txt (accessed on March 2, 2013).
41 C.V. Raman. "The Color of the Sea." *Nature* 108 (1921) 367. Raman was responding here to Rayleigh's works in *Nature* 83 (1910) 48. See also Rayleigh. *Scientific Papers* 5 (1902–1910) 540.
42 Albert Einstein. "Theorie der Opaleszenz von homogenen Flüssigkeiten und Flüssigkeitsgemischen in der Nähe des kritischen Zustandes." *Annalen der Physik* 33 (1910) 1275–1298.
43 C.V. Raman. *Nature.* "Transparency of Liquids and Colour of the Sea." *Nature* 110 (1922) 280.
44 Ibid.
45 Kameshwar Wali. *Chandra: A Biography of S. Chandrasekhar.* Chicago: University of Chicago Press, 1991) 254.

46 RRI archives.

47 h is the Planck's constant, c the speed of light, and m_e is the mass of the electron at rest, θ is the scattering angle. Roger Stuewer. *The Compton Effect: Turning point in physics*. (New York: Science History Publications, 1975) 223–234.

48 Ibid. pp. 249–273.

49 Banerjee. *Physics in Perspective* 16 (2014) 146. https://doi.org/10.1007/s00016-014-0134-8.

50 Ibid. 268.

51 Ibid.

52 Stuewer. (1975) 268. For the argument that though several physicists accepted the Compton effect, but were just as happy to consider light as waves and for the relevance of this in the development of matrix mechanics see Anthony Duncan and Michel Janssen. "On the verge of *Umdeutung* in Minnesota: Van Vleck and the correspondence principle (Part One)." *Archive for History of Exact Sciences* 61 (2007) 553–624.

53 C.V. Raman. "A classical derivation of the Raman effect." *Indian Journal of Physics* 3 (1929) 357–369.

54 Marjorie Johnston, ed. *The Cosmos of Arthur Holly Compton*. (New York: Knopf, 1967) 37. This is a valuable resource which contains Compton's "Personal Reminiscences," a selection of his writings on scientific and non-scientific subjects, and a bibliography of his scientific writings.

55 For example, benzene, glycerin.

56 C.V. Raman. Presidential address to the Indian Science Congress, 1929. (IACS archives).

57 D.C.V Mallik and S. Chatterjee. *Kariamanikkam Srinivasa Krishnan: His Life and Work*. (Hyderabad: Universities Press, 2012) 81.

58 Banerjee, S. *Physics in Perspective* 16 (2014) 146. https://doi.org/10.1007/s00016-014-0134-8 (accessed November, 2018).

59 Ibid.

60 Ibid.

61 Adolf Smekal. "Zur Quantentheorie der Dispersion." *Die Naturwissenschaften* 11 (1923) 873–875.

62 RRI Archives Digital Repository, Bangalore. http://hdl.handle.net/2289/3430 (accessed October 4, 2012).

63 G. H. Keswani, *Raman and His Effect* (New Delhi: National Book Trust of India, 1980) 44.

64 RRI Archives, Doc 17.

65 RRI Archives Digital Repository, Bangalore. http://hdl.handle.net/2289/3430 (accessed October 4, 2012). The portions italicized are mine, to emphasize the point about wave theory.

66 Ibid. pp. 396. For the reference to Bohr's 1913 paper see Bohr. (1913) 1–25. The portion italicized is mine to emphasize the point that Raman was aware of the work of Kramers and Heisenberg. N. Bohr. "On the constitution of atoms and molecules. Part I." *Philosophical Magazine* 26 (1913) 1–25.

67 Singh. *Raman* 409.

68 Sur. *Aesthetics* 25–49.

69 Thomas Kuhn's 1980 videotaped lecture at Harvard entitled "The Crisis of the Old Quantum Theory, 1922–25." I thank Michel Janssen at the University of Minnesota for giving me access to this videotape.

70 Van Vleck. (1926) Vol. 10, part 4 as quoted in Duncan, Janssen. "Umdeutung" for a thorough treatment of the alternative dispersion theories in this period.

71 Charles Galton Darwin. "A Quantum Theory of Optical Dispersion." *Nature* 110 (1922) 841–842.

72 K.F. Herzfeld. "Versuch einer quantenhaften Deutung der Dispersion." *Zeitschrift für Physik* 23 (1924) 341–360.

73 J.H. Van Vleck. *Quantum Principles and Line Spectra*. 10, Part 4 (Washington, DC: Bulletin of the National Research Council, 1926).

74 Jagadish Mehra and Helmut Rechenberg. *The Historical Development of Quantum Theory: 1982–2001, Vol. 6*. (New York, Berlin: Springer, 2001) 354.

75 Ibid.

76 Ramdas was also the first to photograph the scattered spectrum successfully as noted by R.S. Krishnan and R.K. Shankar. "Raman Effect: History of the Discovery." *Journal of Raman Spectroscopy* 10, 198 (1981) 1–8.

77 Duncan and Janssen. "*Umdeutung*" 581–582.

78 Schrödinger. (1926h). In Mehra, Rechenberg. *Historical Development* 121–122.

79 Kramers and Heisenberg. (925) 681–707. It is also, important to note the role of Bohr in Kramers and Heisenberg's attachment with the wave theory. H.A. Kramers and W. Heisenberg. "Über die Streuung von Strahlung durch Atome." *Zeitschrift für Physik* 31 (1925) 681–707.

80 A.G. Shenstone. Rudolf Walther Ladenburg. In Charles Gillispie (ed.). *Dictionary of Scientific Biography* Vol. VII. (New York: Charles Scribner's Sons, 1973) 552–556.

81 Rudolf Ladenburg. "Die quantentheoretische Deutung der Zahl der Dispersionselektronen." *Zeitschrift für Physik* 4 (1921) 451–468. Page references are to English translation in Van der Waerden. (1968) 139–157.

  Also see Duncan, Janssen. "*Umdeutung*" 583 for the above argument of Janssen and Duncan.

82 Van Vleck. (1924) 344, eq. 17. See Van der Waerden. *Sources* 203–222. Here I am following the notation of Van Vleck. I thank Michel Janssen to give me permission to produce these equations. Also see Marta Jordi Taltavull. "The Uncertain Limits Between Classical and Quantum Physics: Optical Dispersion and Bohr's Atomic Model." *Annalen der Physik* 530, no. 8 (2018) 1800104; Marta Jordi Taltavull. "Transformation of Optical Knowledge from 1870 to 1925: Optical Dispersion between Classical and Quantum Physics" (PhD thesis, HU Berlin, 2017); Marta Jordi Taltavull. "Transmitting Knowledge Across Divides: Optical Dispersion From Classical to Quantum Physics." *Historical Studies in the Natural Sciences* 46, 3 (2016) 313–359.

83 Sir Chandrasekhara V. Raman. *The New Physics: Talks on Aspects of Science*. (Freeport, NY: Books for Libraries Press, 1951) Introduction.

84 IACS archives, Kolkata. Raman Correspondence file.

85 Rajinder Singh. "Raman." See also Singh. "Seventy Years Ago: The Discovery of the Raman Effect as Seen from German Physicists." *Current Science* 74 (1998) 1112–1115.

86 Banerjee. "Colonial Physics."

87 Stephen M. Stigler. "Stigler's Law of Eponymy." In *Science and Social Structure: A Festschrift for Robert K. Merton, Transactions of The New York Academy of Sciences*, Series II, 39 (New York: The New York Academy of Sciences, 1980) 147–157. Also see, Robert K. Merton. "Priorities in Scientific Discovery [1957]," reprinted in Norman W. Storer (ed.). *The Sociology of Science: Theoretical and Empirical Investigations*. (Chicago and London: The University of Chicago Press, 1973) 286–324.

88 Suman Seth. *Crafting the Quantum: Arnold Sommerfeld and the Practice of Theory, 1890–1926*. (Cambridge, MA: MIT Press, 2010). Also see Banerjee. "Colonial Physics."

89 Robert Friedman. *The Politics of Excellence: Behind the Nobel Prize in Science* (New York: Henry Holt and Company, 2001) 271.

90 I.L. Fabelinskii. "The Discovery of Combination Scattering of Light in Russia and India." *Physics-Uspekhi* 46 (2003) 1105–1112.

91 The son of J.W. Strutt (better known as Lord Rayleigh).

92 Max Born and Emil Wolf. *Principles of Optics: Electromagnetic Theory of Propagation, Interference and Diffraction of Light.* (London: Pergamon Press, 1959) 1101.

93 Ibid. 1106.

94 Bill Evenson (forum chair for the author's APS, March 2007 talk) in a personal communication with the author remarks that Mandelstam and Landsberg wanted to reflect on their results as to whether they had any more fundamental implications, as opposed to publishing it very quickly, like Raman. To this the author wants to add that this reflection might have been due to the Russians' unawareness of the work of Smekal and Kramers– Heisenberg.

95 I.L. Fabelinskii. *Optika i Spectroscopiya* 55 (1983) 591.

96 Charles Galton Darwin. "The Sixth Congress of Russian Physicists." *Nature* 122 (1928) 630.

97 Max Born. "Fourth Russian Physicists Conference." *Naturwissenschaften* 16 (1928) 741.

98 A. Jayaraman and A.K. Ramdas. *Physics Today* 41 (1988) 56.

99 See internet resource www.iisc.ernet.in/~currsci/nov10/articles33.htm (accessed on May 16, 2007).

100 RRI Archives (accessed May 20, 2019). http://dspace.rri.res.in/jspui/bitstream /2289/5634/1/C%20V%20Raman%20%26%20the%20story%20of%20Nobel% 20Prize.pdf.

101 IISc Archives, see internet resource www.iisc.ernet.in/~currsci/nov10/articles33.htm (accessed May 16, 2007).

102 For Stigler's Law's applicability to kinetic theory and thermodynamics see 1999 paper by John Crepeau in *Physics in Perspective* on Loschmidt, Stefan and Avogadro. John Crepeau. "Loschmidt, Stefan and Stigler's Law of Eponymy." *Physics in Perspective* 11, 4 (2009) 357–378.

103 Credit: Niels Bohr Archive (NBA). I thank Felicity Pors at the NBA for giving me access to this picture.

104 Singh. "Raman" 1489–1494. Sommerfeld was in the United States for Compton's discovery and coined the name *Compton Effect,* for what otherwise might have been called Debye effect or Compton–Debye effect.

105 www.nobelprize.org/nobel_prizes/physics/laureates/1930/press.html   (accessed February 12, 2013).

106 See Banerjee. "Colonial Physics."

107 Arnold Sommerfeld. "Indische Reiseeindrücke." *Zeitwende* 5 (1929) 289–298.

108 Joos to Sommerfeld, May 14, 1928 (Deutsches Museum München archives). See http://sommerfeld.userweb.mwn.de/PersDat/02201.html (accessed May 2012).

109 Rajinder Singh. "Arnold Sommerfeld: The supporter of Indian physics in Germany." *Current Science* 81 (2001) 1489–1494.

110 Rajinder Singh and Falk Reiss. "Seventy Year Ago: The Discovery of the Raman Effect as Seen from German Physicists." *Current Science* 74, 12 (1998) 1112–1115.

111 S. Ramaseshan. "The Portrait of a Scientist—C. V. Raman." *Current Science* 57 (1988) 1207–1220.

112 Edward Said. *Orientalism* (New York: Vintage, 1979). See also Gyan Prakash. "Writing Post-Orientalist Histories of the Third World: Perspectives from Indian Historiography." In Vinayak Chaturvedi (ed.). *Mapping Subaltern Studies and the Postcolonial.* (London: Verso, 2000) 163–190.

113 RRI Archives, Doc 2289–270 (accessed April 10, 2014).

114 Said argues that these stereotypes confirm the necessity of colonial government by asserting the positional superiority of the West over the East. See Said. *Orientalism* (ref. 97), 35, and Leela Gandhi. *Postcolonial Theory: A Critical Introduction.* (New York: Columbia University Press, 1998), 74–80; Richard G. Fox. "East of Said." In Michael

Sprinker (ed.). *Edward Said: A Critical Reader.* (New York: Wiley-Blackwell, 1993) 146–151. The example of "affirmative orientalism" that Fox uses is Indian nationalist leader Mahatma Gandhi's cultural nationalism.

115   Ibid. Fox. "East of Said."

116   Singh. "Arnold Sommerfeld" 1489–1494. Not to be confused with Romesh Chandra Majumdar, the eminent Indian historian.

117   During this time at IISc, Born got into a controversy with Raman over lattice dynamics. For in-depth analysis of the Raman–Born controversy, see Sur. "Aesthetics" 25–49. See also Banerjee. "Colonial Physics."

118   During this time at IISc, Max Born got into a controversy with Raman over Lattice Dynamics. For an in-depth analysis of the Raman–Born controversy see Sur "Aesthetics" 25–49. Max Born to Ernest Rutherford, 22 Oct. 1936, Ernest Rutherford Papers, Rutherford–Born Correspondence, Add. 7653: B297–B306, Cambridge University Library. See also: Abha Sur, "Aesthetics" in *Isis*, Vol. 90, No. 1 (Mar. 1999) 25–49.

119   Parameswaran. *Raman* 106.

120   Wali. *Chandra* 253.

121   Singh. "*Raman*" 1157–1158.

122   IACS archives Folder 3A: undated document on the birth centenary lecture by Ramaseshan on Raman in 1988 and Silver Jubilee of the Raman Effect held at IACS Calcutta.

123   Fabilinsky. "The discovery" 1105–1112.

124   Sur. "Aesthetics" 46.

125   The difference between Raman's nationalism and that of Bose and Saha can be viewed as part of a larger theme of how Indian nationalism played out regionally, for example in Bengal versus that in South India.

126   Here I mean there is a distinction between nationalism and anticolonialism, which are subtly different. See Ranajit Guha. *A Subaltern Studies Reader, 1986–1995* (Minneapolis: University of Minnesota Press 1997) 35–44.

127   Max Weber. *Essays in Sociology.* (New York: Oxford University Press, 1946) 129–156.

5 Meghnad Saha

Applying the light quantum

In the introduction to his well-known *Theoretical Astrophysics: Atomic Theory and the Analysis of Stellar Atmospheres and Envelopes*, Norwegian astrophysicist Svein Rosseland remarked on the importance of Meghnad Saha's contributions:

> Although Bohr must thus be considered the pioneer in the field [atomic theory], it was the Indian physicist Meghnad Saha who (1920) first attempted to develop a consistent theory of the spectral sequence of the stars from the point of view of atomic theory. Saha's work is in fact the theoretical formulation of Lockyer's view along modern lines, and from that time the idea that the spectral sequence indicates a progressive transmutation of the elements has been definitely abandoned. From that time dates the hope that a thorough analysis of stellar spectra will afford complete information about the state of the stellar atmospheres, not only as regards the chemical composition, but also as regards the temperature and various deviations from a state of thermal equilibrium, the density distribution of the various elements, the value of gravity in the atmosphere and its state of motion. The impetus given to astrophysics by Saha's work can scarcely be overestimated, as nearly all later progress in this field has been influenced by it and much of the subsequent work has the character of refinements of Saha's ideas.[1]

Meghnad Saha (1893–1956), an eminent *bhadralok* scientist with a lower caste background and born in a remote Indian village, played a key role in developing the theory of thermal ionization, and its application for explaining stellar spectra using thermodynamics and kinetic theory of gases in the 1920s. The Saha equation, as it is now known, originated from Saha's insights from working in Calcutta. The ideas developed by Saha were first given in the paper, "On Ionization in the Solar Chromosphere," published in the *Philosophical Magazine* (1920). The social dimensions of Saha's life played an important role in shaping his ideas in science.[2]

Saha belonged to the lowest strata of Indian society, and suffered from its associated disadvantages. The grip of this traditional caste system is still very firm in India. The caste status is bestowed upon the individual at birth. Therefore, it may be termed as an "ascribed" status, which should not to be confused with the "achieved" status.[3] Saha was born as a *shudra*, which had the lowest position in the caste hierarchy. Consequently, he was ascribed to the lowest status at birth as previously stated, and he was deprived of the opportunities available to upper caste members. He faced a double power differential in colonial India—as a scientist under British rule, and as a marginalized person in his own society because of his lower caste (*shudra*) status.

There are two main arguments in this chapter. The first argument is about Saha's nationalist aspirations and involvement with Bengal Revolutionary groups such as Jugantar, Anushilan Samiti, and the Bengal Volunteers, and how he saw his work facilitating the process of decolonization. The second is about how Saha, from his humble *shudra* origin, raised himself from a lower caste to the status of a *bhadralok* through his significant scientific contributions. Interestingly enough, such a transition from the lowest caste to the prominent *bhadralok* position did not occur for C.V. Raman and H.J. Bhabha, who were born into Brahmin families, and therefore into the topmost rank in the Indian caste hierarchy. Satyendranath Bose was born as *Kayastha*, a caste that is a type of middle class. Moreover, his family was active in bureaucracy. However, Bose statistics forged a new identity in Bose's social life, by virtue of which he was identified as a *bhadralok* scientist and not a lower *Kayastha*.

Saha was closely associated with the Bengal Revolutionaries, especially with Anushilan Samiti, Jugantar, Bengal Volunteers, Jatindra Nath Mukherjee (commonly known as Bagha Jatin), and Pulin Das. Their rationale was to put up an armed resistance against the British rule for decolonization. Because of this association, Saha's early life was full of hindrances imposed by the colonial government. Hence, Saha faced a two-fold problem—of being a *shudra* in the first place, and being a revolutionary early in his life—continuously finding himself on the wrong side of "law," and constantly being hounded by the British secret services.

Though the exact nature of Saha's involvement with the revolutionaries is not fully known, it is worth noting that, to be closely involved with the revolutionary movement was rather unusual in those days for an Indian scientist. The life of Saha therefore reflects a different dynamic, not found in the life of most of his contemporaries. Despite the difficulties he faced from his involvement in the revolutionary movement, Saha managed to establish himself as a professional *bhadralok* scientist, making everlasting contributions to physics and never hesitating to collaborate with Western scientists. This type of synergy of forging international collaboration, while being grounded in the situation of the Indian nation was another characteristic of *bhadralok* physics. Typically, such a pattern of intellectual flexibility was not seen with fellow scientists Homi Bhabha and Ganesh Prasad.

Organizing science for the creation of an independent modern nation was the overarching theme that Saha pursued actively. He was trained in India early in the twentieth century, and he combined the pursuit of science with the rising tide of nationalist aspirations. He considered the development of modern scientific research and its institutions in India as an essential component of acquiring

national independence. Using Saha's early career trajectory and professionaliza-
tion, this chapter also outlines how science was practiced in early-twentieth-century India. Saha's dialogues with Jawaharlal Nehru and Mahatma Gandhi are also
discussed, particularly the nature of their diverging paths to forming the Indian
nation-state. More importantly, this chapter examines how Saha disagreed with
Nehru and Gandhi on several occasions, especially in the context of development,
and the role of science in the development of the Indian nation.

The transition from *shudra* to *bhadralok*

Meghnad Saha was born in 1893, as one among eight children of a poor, low-caste
shopkeeper in the town of Seoratali in East Bengal (present day Bangladesh).[4] He
was the fifth child of his parents, Jagannath Saha and Bhubaneswari Devi. Saha's
elder brother had failed in high school. So, his father decided that Meghnad would
work in the family's shop selling groceries, just as his elder brother did. It was
Meghnad's mother and uncle who intervened and allowed Meghnad to continue
his high school education.[5]

Since Seoratali did not have a proper middle-school, with the nearest school
being in a distant village, Ananta Kumar Das, a local medical practitioner and
a *kaviraj* of the *ayurvedic* tradition,[6] agreed to help Saha by providing him free
housing and a stipend because of Saha's unusual intellect. In return, Saha agreed
to wash his own dishes and assist Das in other household chores.[7] In 1905, having finished middle-school, and ranking first in class at the age of twelve, he
was awarded a scholarship to study at the Dacca Collegiate School. But he was
expelled soon after admission because of his participation in a protest rally organized by fellow students. Together with some other senior students, Nil Ratan Dhar
(Saha's close friend who was earning his B.Sc. in Chemistry at the time and who
would later go on to form a school of chemistry at Allahabad University) among
them, Saha and his peers took off their shoes, as a sign of disrespect, and staged
a boycott during a school visit by the Bengal Governor, Andrew Fraser, in order
to protest against the Partition of Bengal earlier that year under the Viceroyalty
of Lord Curzon.

The Partition of Bengal sparked the Indian nationalist movement in ways that
even the founding members of the Indian National Congress had not envisioned.
Having lost his scholarship, Saha joined Kishori Lal Jubilee School and passed
the entrance examination for Calcutta University in 1909, standing first among
thousands of students from the schools of East Bengal. His explorations in science, as Robert Anderson argues,[8] began at the age of sixteen in Dacca College
in 1909, where his teachers were E.C. Watson in chemistry, B.N. Das in physics,
and N.C. Ghosh as well as K.P. Basu in mathematics. Saha also began learning
the German language from the Austrian scientist P.J. Brühl, who taught at Bengal
Engineering College.

In 1911, Saha cleared the Intermediate Science Examination of Calcutta
University from the Dacca College. He ranked first in physics and mathematics,
but third in the whole examination. While studying at Presidency College from

1911 to 1913, Saha stayed in the Eden Hindu Hostel where he had to go through the ordeal of casteism. Some students objected to eating in the same dining hall with Saha, because he belonged to the lowest caste. He was also prevented by some Brahmins (the highest caste) from making a religious offering to the goddess of learning, Mother Saraswati.[9]

At the age of eighteen, Saha began his B.Sc. in mixed mathematics at Presidency College in Calcutta, where he earned the nickname "Eigenschaften" for his seeming "invincibility" and his knowledge of German.[10] Saha and Satyendranath Bose were classmates at Presidency College where the well-known chemist and entrepreneur Prafulla Chandra Ray, the internationally acclaimed physicist cum plant physiologist Jagadish Chandra Bose, and the famous mathematician D.N. Mallik were among the teaching faculty. The teachers inspired their students to use science as a tool for promoting the spirit of nationalism, e.g., Jagadish Chandra Bose, in his book *Response*, epitomized the linking of nationalism and science and dedicated the book to the people of India saying, "To my countrymen, who will claim the intellectual heritage of their ancestors."[11]

For Saha, the pursuit of science and aspirations for national independence were linked together even more closely. During his studies at college, he encountered the militant nationalists of Bengal, including Subhas Chandra Bose, Sailen Ghosh, and many others. Because of the clandestine nature of their activities, the extent of Saha's involvement with them has not yet been thoroughly examined. Bengali revolutionaries drew much of their inspiration from a parallel struggle for independence in Ireland and the Irish revolutionary organization called the *Sinn Fein*, which became one of the most important models for militant nationalists in Bengal.[12]

Their admiration for Irish nationalism and emulation of Irish tactics demonstrated divergence of strategies within Indian nationalism. The Bengali nationalism in the eastern part of India, with its focus on violent method of insurrection, differed from what was commonly perceived as the elitist nationalism led by Gandhi and Nehru, with their avowed policy of non-violence. As Irish people were also impacted by British colonialism, many Indian nationalists found a cultural similarity with Ireland.

Saha became involved with the Bengal revolutionaries who strove to procure arms from abroad, raise funds needed for campaigns, inform people at home and abroad about the plight of Indians under colonial rule, recruit new workers, especially from people belonging to the *bhadralok* category, and offer shelter to absconders.[13] Saha joined *Anushilan Samiti*, a revolutionary organization that had close ties with the *Ghadr*[14] movement organized by the Indian revolutionaries abroad in San Francisco and Canada. A few members of the *Ghadr* party based in Berlin informed Bagha Jatin (Saha's friend in college) that Kaiser Wilhelm II was sympathetic toward the freedom fighters of India. Consequently, Berlin was willing to supply arms for them.

Secret plans were also chalked out for the purchase of arms sent to India from the United States. A joint Indo-German organization known as the Indian Revolutionary Committee was set up in Berlin to coordinate the efforts of the

Indian revolutionaries that were globally scattered. Transporting arms in a secretive fashion was quite hazardous in those days, because of the strict surveillance by the British police.[15] Bagha Jatin was informed that the Kaiser was sending weapons in a ship (coming from Singapore) for the *Anushilan Samiti*, and Saha was assigned the task of picking up the weapons from the ship arriving at the coast of the Sunderbans in south Bengal. Their expectations were thwarted. Saha came back from the Sunderbans empty handed, as the ship did not arrive.[16] The colonial government, however, did not take Saha's revolutionary relations lightly, as will be discussed later in the chapter.

It may be pointed out here that Saha never set aside his pursuit of science, even during his participation in the revolutionary activities. He especially wanted to popularize science amongst common people. His first article in Bengali on Halley's Comet came out in the Dacca College Magazine in 1910, and included a remarkably lucid and interesting explanation of the comets. Saha explained the physics within the article clearly, as he did not use any technical jargon. Writing on various aspects of science for the general public became a hobby for Saha, especially in the vernacular language. Later, as he became involved with scientific circles in India, his popular writings stopped, when he started his own journal called *Science and Culture*.[17]

After four years of study in Calcutta, Saha tried to appear for the Indian Finance Service (IFS) examinations in 1915. Though he ranked second in the M.Sc. examination of Calcutta University, he was refused permission to write the IFS on grounds of his association with the revolutionaries.[18] It led to his continuing pursuit of graduate studies in applied mathematics and physics at the newly established University College, Calcutta.

Saha's transition from a graduate student to a physicist of professional stature was a remarkable process in which several unforeseen events, such as transferring to a different department and gaining an opportunity to teach a graduate seminar, helped him to become acquainted with the current state of research in theoretical physics. Although a limited amount of European scientific literature and very few advanced books were available in Calcutta libraries as a result of the World War I, Saha obtained effective help from P.J. Brühl, the Austrian scientist in the Bengal Engineering College. Brühl possessed a good collection of advanced texts and journals of physics in German language, including a rich repertoire of papers on quantum theory and relativity.

To appreciate the issue in perspective, the role of some eminent personalities of that time needs to be mentioned. Sir Ashutosh Mukherjee, a famous mathematician who was also the vice-chancellor of Calcutta University from 1906 to 1914, invited Saha to work as a lecturer in the department of mathematics at the newly opened University College of Science for post-graduate studies and research in Calcutta. The establishment of this new college became possible due to the endowments made by two front-ranking lawyers of Calcutta—Tarak Nath Palit and Rash Behari Ghosh. Because of subsequent conflicts with Ganesh Prasad, the head of the mathematics department, Saha (along with Satyendranath Bose) was transferred to the physics department.

Unlike the mathematics department, the physics department had a dearth of teaching faculty, which compelled Saha to teach a graduate seminar on thermodynamics and spectroscopy. Saha also lectured in postgraduate classes on hydrostatics, the figure of the earth, and was also in charge of the heat laboratory. Teaching thermodynamics and spectroscopy was a challenge as these topics were new to him, but that did not deter him from learning the subjects in an immaculate fashion. While teaching, Saha read *A History of Hindu Chemistry* by his mentor and eminent Indian chemist Prafulla Chandra Ray. He also read Planck's *Thermodynamics* and Nernst's *Der neue Wärmesatz*, and he familiarized himself with the papers of Bohr and Sommerfeld on the quantum theory of the atom.

Therefore, Saha became aware of the current state of research in thermodynamics and spectroscopy, and developed a solid foundation in classical physics. At the age of twenty-four he published his first paper in the British scientific journal, *The Philosophical Magazine*, on the theory of Maxwell's electromagnetic stress-energy tensor. He continued with research in electrodynamics by deriving the Lenard–Wiechert potential due to a point charge, and calculated the radiation pressure of light in 1918.[19] He also came across a work by the nineteenth-century Irish astronomer Agnes Clerke on the Sun and the stars. This research gave him the necessary background for further explorations in astrophysics.[20]

On November 12, 1919, *The Statesman*, a daily newspaper published in Calcutta, sent a correspondent to the astronomical observatory at the Science College campus to get an explanation of a cabled confirmation from the Reuters, on Einstein's prediction of deflection of starlight in the gravitational field of the sun.[21] This cabled message was published in *The Statesman* on the same day and stated:

London, Nov 7

An announcement made at the Royal Society, which is described in the Press as overthrowing the certainty of ages, and requiring a new philosophy of the universe has aroused intense interest in scientific circles in view of its all important bearing on the fundamental physical problem. Sir Frank Dyson, Astronomer Royal, expressed the conviction that the results of recent experiments were definite and conclusive; that light from the stars as it passed the sun was deflected owing to the presence of the sun, this deflection closely according with the theoretical degree, predicted by Professor Einstein, namely that the deflection was twice the amount laid down by Newton. The discussion which followed was very intricate, no speaker succeeding in giving a clear non- mathematical statement. The results of the experiments were generally accepted, but the theoretical bearings provoked much debate.[22]

In response to this report, a news correspondent of *The Statesman*[23] was instructed to contact Meghnad Saha (as he was translating Einstein's *Special Theory of Relativity* into English), and meet him in the astronomical observatory of the Presidency College. On request, Saha promptly wrote a popular explanation of this effect and gave it to the reporter. This explanation was published in the same newspaper the following day on November 13, titled "Time and Space—The New Scientific Theory." Saha's *Statesman* entry started as follows:

> The announcement conveyed in yesterday's Reuters's cable that Professor Einstein's theory of the equivalence of Time and Space has at last been veri-fied by observations made during the last solar eclipse, will be hailed with joy by scientific circles all over the world. If the announcement be true, then the time-honour dogma, that time and space are quite independent of each other, will be subverted once for all ... The new theory is thus of great interest to astronomers and the physicists from the point of view of absolute measure-ments. It will prove to be of interest to those physicists who are trying to unravel the inner constitution of the atom with the aid of dynamics. But for the measurements in the solar system, nothing appreciable is to be expected.[24]

Beginning his popular exposition on relativity by giving a brief history of the topic, Saha summarized the theoretical and experimental research on relativity of vari-ous physicists, including Hendrik Lorentz, Hermann Minkowski, Hermann Weyl, Willem de Sitter, Arthur Eddington, Albert Michelson, and Edward Morley. Saha concluded his expose of Einstein's new theory of space and time with an explana-tion of the perihelion precession of Mercury.

He also showed how predictions of Einstein's theory of light bending was verified without doubt by Eddington during the solar eclipse of 1919, when bright stars were visible for a few minutes in the duration of the eclipse. Though Saha gave a popular explanation in the newspaper for non-experts interested in scien-tific matters, there were some critical comments about his explanation that were not atypical at the time, because of the revolutionary nature of Einstein's theory. On November 14, the newspaper published a letter signed as Simplicimus, with complaints about the counterintuitive nature of the new science:

> Dr. Saha's contribution to this morning's issue of the Statesman on "The New Scientific Theory" of time and space appears to be intended for the non-expert who is sufficiently interested in such matters to wish to know. That is my case, but I confess myself still completely at a loss. This "relativity of space" is evidently something not metaphysical at all, not the Kantian "form of thought", but physical; apparently, on this view, an inch is not always an inch but more or less according to circumstances. Obviously I am talking nonsense, which may be regarded as the pathological effect of Reuter's and Dr. Saha's explanations of things on an average mind. Can nobody help?[25]

Saha responded to this letter, attempting to explain some of the general philoso-phy of relativity theory, while agreeing that it was a challenging task to give this philosophy a popular explanation. In his explanation Saha remarked:

> Apropos of Simplicimus's letter to this morning's issue of The Statesman, I wish to add the following lines which I hope may make certain passages

of Reuter's telegraph (especially the words 'certainty of ages overthrown', etc.) clearer. It will not be correct to say that Newton's law of inverse square is false. The real point at issue "what is meant by the mass of a body, or the distance between two particles?" It will not do for precise astronomical purpose, as we have hitherto done, to take a standard rod, and find out how many times this is contained between two particles, but we must go deeper into the conception of time and space ... it is not possible to give a popular idea of the theory which even the savants of the Royal Society found rather exacting.[26]

Trying to succeed in giving popular explanations, Saha and his colleague, Bose, translated Einstein's papers on special and general relativity, which the University of Calcutta Press later published in 1920 as a book titled *Principles of Relativity*. This was the first translation of Einstein's work in English. Studying relativity also gave Saha the opportunity to read up on several facets of classical physics, especially electromagnetic theory.

Many people inside and outside of the scientific community began to acknowledge Saha's pursuit of science. In 1919, he was awarded the prestigious Premchand Roychand Scholarship from the University of Calcutta. He used the scholarship to do research at the University of London (under British astrophysicist Alfred Fowler), and in Berlin where he worked in the laboratory of German physicist Walther Nernst in 1921. The letter of reference from Ashutosh Mukherjee greatly helped Saha to obtain a research leave for Europe. During that time, Mukherjee was the President of the Post Graduate Council of Teaching in the Calcutta University, and he gave a glowing certificate of recommendation on August 20 to the Council of Post Graduate Teaching (Senate House, Calcutta), summarizing in brief Saha's academic trajectory till 1921:

Dr. Meghnad Saha has been personally known to me for many years as a brilliant and devoted student of the University. His attainments have been of a very varied character as will appear from the following statement:

Passed the Entrance Examination in 1909; stood first in Eastern Bengal, first in Mathematics, first in Literature (English, Bengali and Sanskrit) in Eastern Bengal. Passed the I.Sc. Examination with German as additional subject; stood third, first in Mathematics and Chemistry.

B.Sc. in 1913 with First Class Honours in Mathematics; stood second; Hindu college Foundation scholar.

M.Sc. in 1915 in Mixed Mathematics (Class I, stood second). D.Sc. in 1919 in Experimental and Theoretical Physics.

Premchand Roychand Scholar in 1919 in Physics.

Works – Author of "Principle of Relativity" containing translations of original papers of Einstein and Minkowski.

Appointed University Lecturer in Physics and Applied Mathematics in 1916. Lectured to the post-graduate students on the following subjects:

Advanced Heat including Thermodynamics and Kinetic Theory	1916–20
Radiation and Quantum Theory	1916–20
Electro and Magneto-Optics	1919–20
Hydrostatics	1916–18
Figure of the Earth	1918–20

He has performed his duties as University Teacher to our entire satisfaction and I am not without hope that after his return from Europe he will be able to devote himself to research and to undertake the promotion of research amongst students, with increased zeal and efficiency. I shall watch his work in foreign countries with continued interest, coupled with hope that he may fully maintain the reputation of his Alma Mater.[27]

Ashutosh Mukherjee, himself a Brahmin (the highest tier in the Indian caste system), gave Saha opportunities because of his merit—an attribute of a *bhadralok*—and academic accomplishments—additional attributes of a *bhadralok*. Before gradually acquiring the status of a *bhadralok*—a well-mannered individual having talent, academic accomplishments, and a local and global patronage network—Saha had to suffer in his college life because of his lower caste background. By means of his firsthand experience, Saha addressed this question of caste when he commented:

It is a standing complaint that, at present time, the hostels that are attached to colleges are practically the monopoly of a few aristocratic classes—viz., of the Brahmins, the Kayasthas, the Vaidyas and the Nabasaks. Members of democratic classes are either not admitted, or if at all admitted, they are allowed to live not as a matter of right, but as a matter of grace. If any student of the orthodox type demurs to living with them in the same room, and taking meals in the same dining hall, the unfortunate student is asked to remove to some other place, and take his meals in his own room. The writer knows of several instances where this state of things has actually existed. Now the members of the lower castes feel that, in at least these hostels which have been constructed at public expense, they have the same rights as members of other castes. They expect that they should be admitted freely and allowed to live in a manner consistent with their ideas of self respect and dignity. It will not do if separate hostels are opened for them, for in that case, at least twenty-five separate communal hostels should be opened for each college, one for the use of each particular community. The government, as a matter of principle, does not make any distinction of caste or creed in points of law or employment. The same principle should be adopted in this case. Those students, or communal leaders, who find it irreligious to dine with their fellow-brethren of other castes, should be asked to shift for themselves, or construct hostels at their own expense. They should not be allowed or encouraged to introduce a feeling of discord in the pure academic atmosphere.[28]

This quotation shows that he felt the necessity of changing the prevalent discriminatory practices when he was in college. He suffered during his stay in the college hostel, but never protested against this differential treatment. Then later, when he achieved a higher social status as a *bhadralok* scientist, he reminisced about his early life and expressed his displeasure (as seen in the quotation). He made himself upwardly mobile, and reached the highest strata of society. There was no change in his caste position, since there was no room for mobility in the caste system. But he came to belong to the special class of *bhadralok*, in which vertical mobility[29] was unrestricted by one's caste position. Despite the colonial situation, opportunities were available to him, by means of which he could achieve a new status. His lineage, ethnic membership, and his father's position did not rule out his movement from a lower to a higher class of society.

In the above statement, Saha used the term "democratic classes" instead of the word "subaltern" because in democracy, number matters the most, and people from the lowest caste were and still are numerically stronger than in other castes. Saha acknowledged the fact that the government did not "discriminate" against him, or anyone for that matter, on the ground of caste. The British rulers provided equal opportunities to many Indian scientists, which was an important democratic principle. Saha took advantage of the opportunities available for him in order to raise his social status. These opportunities led him to the acquisition of higher education in science and, subsequently, the publication of articles in journals and periodicals, namely *Philosophical Magazine*, *Physical Review*, *Journal of the Asiatic Society of Bengal*, and *Astrophysical Journal* among others. These achievements accorded Saha a higher status in society, and people gradually became familiar with Saha, not just as a scientist but as a *bhadralok* scientist. Though this gradual ascendance into becoming a *bhadralok* was slow, it was also fruitful for both Saha and Indian science as a whole, considering that now anybody can work in the field of science.

MIT historian Abha Sur has advanced the view of "democratic classes." I disagree with her because her argument (see later) is overstated, especially regarding caste discrimination. Saha did face caste discrimination early in his life before becoming a *bhadralok* scientist. However, Sur argues that all his life Saha was bothered and discriminated against because of his lower caste status. I argue that, in becoming a *bhadralok* scientist, Saha transcended this narrow caste status, and his colleagues (both Indian and British) gradually saw him through the lens of his science and not his caste. My argument does not reduce Indian science to caste politics in colonial India, unlike Sur's argument. Caste was and still is a problem in India. But scholarship should gradually acknowledge the agency of Indian intellectuals who made achievements despite coming from a lower caste, instead of being fixated on the problems caste politics entailed.[30]

Saha's scientific pursuits

In 1919, when Saha was a professor of physics in Calcutta University, he published his first important work in astrophysics containing the formulation of the concept of selective radiation pressure, and the recognition of its role in the relative distribution of elements in the solar atmosphere.[31] He attempted to examine the contradictions that emerged when one tried to explain the tail of comets and other astrophysical phenomena (such as solarprominences, corona) using the wave theory of light. Previous works by Karl Schwarzschild, Seth Barnes Nicholson, and Otto Julius Klotz concluded that for particles of molecular size (radius = 10^{-8} cm) the effect of light pressure is totally evanescent. But this result seemed rather contradictory to Saha as it did not conform to the requirements of astrophysics.[32]

Saha's paper "On Radiation Pressure and the Quantum Theory," published in the *Astrophysical Journal* in 1919, used quantum theory to explain repulsive forces[33] (levity) acting on gaseous molecules, and reducing the gravitational attraction on them. As had been shown experimentally by Lebedew,[34] radiation pressure acts upon the molecules of absorbing gases like carbon dioxide, methane, and propane, among others. Saha argued that despite the failure of the wave theory to account for this occurrence, molecules do experience a radiation pressure which conforms to Maxwell's law:

> After the prediction by Maxwell of the existence of the pressure of radiant energy, on the basis of his theory of stresses and strains in ether, other ways of arriving at the same result have been found by Bartoli (thermodynamical), Poynting (flow of momentum along a ray of light), and Larmor (electromagnetic wave-theory of light). A review of these methods shows that they are all statistical ... Schwarzschild and more recently Nicholson and Klotz have worked out, on the basis of continuous theory, the value of the radiation pressure, when the size of the obstructing mass is gradually decreased, ultimately being reduced to the scale of the wave-length of light. It appears from these investigations that for particles of the molecular size (radius = 10^{-8} cm) the effect of light-pressure is totally evanescent. But this conclusion from the old continuous theory is rather contradictory to the requirements of astrophysics, for in order to explain tails of comets and other astrophysical phenomena (such as solar prominences, corona) which take place on the surface of luminous heavenly bodies we have to assume the existence of certain repulsive forces (levity) acting on the ultimate gaseous molecules and thus reducing the gravitational attraction on them ... It may thus be taken for granted, in spite of the failure of the continuous theory, that the molecules do really suffer a radiation pressure.[35]

Saha went on to explain radiation pressure without using explanations like electrical forces[36], or the suggestion that gravitational force might decrease with temperature,[37] but instead, by using the quantum theory of Einstein and Bohr. Following quantum ideas, Saha assumed that light is localized in pulses (of hydrogen light corresponding to the line Hα by the hydrogen atom) having energy hv, encountering a molecule m and being absorbed by it. The molecule's forward momentum would be (hv/c) and mass (hv/c^2), and the velocity of the molecule would be (hv/mc). Using h = 6.54×10^{-21}, $\lambda = 6.563 \times 10^{-5}$ cm; and m = 1/(6.062×10^{23})

gms. Saha calculated the impulsive velocity of the molecule as 60 cm per second at each "kick" of light. Notable here is how the movement imparted to light is framed by Saha as a "kick."

While this velocity is small compared to the orbital velocity of the molecules, the total velocity acquired by a hydrogen atom per second will be contingent upon the number of "kicks" of light it experiences per second. Provided that these "kicks" of light are sufficiently great, the velocity acquired may rise to greater values which, in turn, explain radiation pressure and Lebedew's results.[38]

Saha's formulation developed well before the Compton Effect was defined— a moment that is argued as a "turning point" in physics.[39] Though physicists in Europe were skeptical about the light quantum before the Compton Effect, Saha used light-quantum very effectively (unlike Raman) in developing ideas that can be applied to explain astrophysical phenomena like tails of comets. Furthermore, Saha argued that while according to the continuous theory, the atom must be illuminated for at least fifteen minutes before it can acquire sufficient energy for the emission of the electron, the emission actually takes place in less than 1/1600 of a second after illumination. Using this notion, Saha explained that the tail of comets is caused by some sort of repulsive action exerted by solar light. Quite contrary to this concept, the continuous theory found that the effect was evanescent on particles of the molecular size, and that the tail was supposed to consist of some sort of cosmic dust to be explored later. Spectroscopic observations of the light from the tail show results to the contrary (i.e., they show that the light comes from partly luminous gases).

Saha further explained this theory by arguing that as the comet approaches the sun, several pulses of light from the sun traverse the nucleus and the nebulous envelope (coma) around the nucleus. Few of these pulses are picked up by the luminous gases, which gradually gain in velocity until they become sufficient for escape from the main mass of the cometary matter that then forms into the tail.

By using quantum theory, Saha explained a basis for the existence of radiation pressure on atoms and molecules, which was also confirmed by Peter Lebedew's experimental results.[40] The radiation pressure is the same for both Lebedew and Saha. Lebedew's initial motivation for studying pressure due to radiation was an attempt to confirm the continuous Maxwellian electrodynamics.[41] Later, his experimental results showed that the Maxwellian continuous theory cannot explain it successfully.[42] On the contrary, Saha's starting point in 1919 was in using the discontinuous nature of light, which entailed applying Planck's law and Einstein's photoelectric effect to explain the repulsive action exerted by the sun's light.

The pressure $p = \Sigma\Sigma\ h\nu = AI$, where the summation is extended over all the pulses absorbed in unit time, within unit area, and I is the intensity of light, while A is the fraction absorbed. Saha had in mind Einstein's 1905 paper on photoelectric effect of the velocity of emission of photoelectrons, with the velocity of escape given by $\Sigma\ \frac{1}{2}\ m\ v^2 = h\nu - \Sigma A$. The maximum velocity occurs when only one electron is emitted. Then the above relation becomes $\frac{1}{2}\ m\ v^2 = h\nu - A$. Saha was confident of this outcome as he was also aware of Millikan's 1916 experiments confirming this law quantitatively, and the fact that it could not be explained by continuous theories of absorption.

This confirmation was also an important advance in physics, because according to the older continuous wave theory, the pressure of light was virtually nonexistent on obstacles at the atomic or molecular size. Saha was not alone in his research findings that contributed to the burgeoning field of "astrophysics." British astronomer Arthur Eddington was a well-known contributor to this field, as he had just verified Einstein's general theory of relativity in 1919. However, Eddington, in a communication to the *Astrophysical Journal*, remarked:

> As there seems to be a rather widespread impression that gases are not subject to radiation-pressure, it may be advisable to state the theory briefly. The pressure is simply a consequence of absorption or scattering. A beam of radiation carries a certain forward- momentum proportional to its intensity, after passing through a sheet of absorbing medium, a weaker beam emerges carrying proportionately less momentum ... The medium, in fact, absorbs the momentum of the beams in the same proportion as it absorbs energy.[43]

While Eddington was also interested in similar astrophysical problems that involved the explanation of radiation pressure, his view differed from Saha's. Eddington considered the aggregate effect of light pressure in the interior of stars (i.e., the region where the gaseous atoms are under such a high pressure that they no longer emit, or absorb, waves of a specific type but of all wavelengths). Meanwhile, Saha considered the atmospheres of luminous bodies, having very low pressure such that gaseous atoms could emit their own characteristic radiation. Based on this work on selective radiation pressure, Saha was awarded the D.Sc. degree from Calcutta University in 1918.[44]

Nationalism

1919 was also a significant year in the history of Indian nationalism. To suppress revolutionary activities in India by extremists in the aftermath of the Great War, the controversial Rowlatt Act was passed by the British government, specifically by the British judge Sir Sydney Rowlatt. The purview of the act was vast. Any person (man or woman) could be arrested without a warrant, and with no provision of appeal. Suspected revolutionaries could be detained without arrest, denied an open trial, and subjected to severe restrictions on their everyday movements.

The Rowlatt Act sparked voices of protest all across India, particularly Bengal and Punjab, where the revolutionary ferment was the most extreme. In the wake of the Rowlatt Act, Gandhi launched the *Satyagraha*—a movement of passive resistance, which included the boycotting of British goods, and promoting indigenous products and strikes by the masses that gained momentum in several towns. The British dealt strongly with the Indian backlash against the Rowlatt Act by imposing martial law wherever possible. On April 13, 1919, the massacre of Jallianwala Bagh took place in Amristar, Punjab, when Brigadier General Dyer opened fire on an unarmed crowd protesting against the deportation of nationalist leaders Satyapal and Kitchlew.

Official sources estimated the number of casualties as 379 killed, including women and children, while unofficial accounts gave significantly higher figures.[45] Saha remarked later how he felt about this tragedy:

> Everyone in our great country who has his own interest and that of his fellow-men at heart must have felt extremely grieved, nay shocked at the diabolical outbreak … leading to violence and large scale massacre of innocent people, which have disfigured the history of this country at this present momentous epoch. These incidents have degraded us before the whole world and have rendered the prospects of a peaceful betterment of conditions of living for the common man in this country an extremely remote one.[46]

Traumatized by these events and by the ongoing violence, Saha took respite in thermodynamics, spectroscopy, quantum theory, and the techniques of physics, which provided him an escape from the depressing situation of colonial India, and gave him the agency to make contributions that benefitted the Indian nation.

In 1919, Saha published a paper in the *American Astrophysical Journal* titled "On Selective Radiation Pressure and its Application." In 1920, he pioneered the "Ionization Theory," which led to his election as a Fellow of the Royal Society in 1927. When attempting to answer a question that had long plagued astronomers and astrophysicists as to why stars exhibited different spectra, Saha used what he had learned in thermodynamics, spectroscopy, and quantum theory while performing his teaching duties at Calcutta University. Saha's work in this field of ionization theory made a great impression in the then newly emerging quantitative field of theoretical astrophysics.[47] Dilip Salwi explains that "These articles were original enough to secure for him the Doctorate of Science."[48]

The general problem Saha tackled was to explain some of the apparent anomalies in the chromosphere spectra of the Sun. These explanations involved explaining the flash spectrum of the chromospheres and why calcium, in spite of being heavier, is present at much greater heights in the chromosphere than hydrogen. British physicist Norman Lockyer's work on solar physics resided in the background, seemingly incomplete.[49] His work suggested that spectral lines originating in the flash spectrum at the top of the chromospheres were enhanced because they had a higher intensity and were more strengthened. Saha remarked:

> It would lead us to the hypothesis that the outer chromosphere is at a substantially higher temperature than the photosphere and the lower chromospheres; and that the temperature of the Sun increases as we pass radially outwards. This hypothesis is, however, quite untenable and is in flagrant contradiction to all accepted theories of physics … A more plausible explanation is that the lines in question are not due to radiations from the normal atom of the element, but from an ionized atom, i.e., one which has lost one electron. The high-level chromosphere is, according to this view, the seat of very intense ionization.[50]

Attributing agency to ionization, and using quantum discontinuity as entailed in Einstein's photoelectric effect, was a new and very suggestive idea in 1920. By doing so, Saha could explain why spectral lines appeared, became enhanced, and then decreased in intensity. While ionization explained this phenomenon, one still had to explain why there was preferential ionization at the upper levels of the solar chromosphere. Subsequently, Saha confronted this problem.

Saha conceptualized ionization of chromospheric elements as a process that depended on temperature and pressure. Furthermore, using the "ideal gas law" and by changing the parameters (temperature and pressure), one could observe the effect on the concentration of ions. But how does one go about writing an equation between temperature, pressure, and concentration of ions? Saha dealt with this problem by using the Berlin physical chemist John Eggert's work in the German journal *Physikaliche Zeitschrift* as a basis.[51] Eggert, who was Walther Nernst's student, viewed ionization as a chemical reaction in which he applied Nernst's Heat Theorem to explain the ionization in stars due to high temperatures. Years later, Saha commented about his reflections:

> I was a regular reader of German journals which had just started coming after four years of the First World War, and in the course of these studies I came across a paper by J. Eggert ... who had given a formula for thermal ioniza-tion, but it is rather strange that he missed the significance of the ionization potential of atoms, the importance of which was apparent ... while reading Eggert's paper I saw at once the importance of introducing the value of the ionization potential in the formula of Eggert, for calculating accurately the ionization, single or multiple, of any particular element under any combina-tion of temperature and pressure. I thus arrived at a formula which now goes by my name [Saha ionization formula].[52]

The Saha equation showed that line strength in a star's spectrum was primar-ily a temperature effect and secondarily a pressure effect, and both were arti-facts of thermal ionization— the loss of electrons by collision with other atoms.[53] Discussing the case of the ionization of Calcium, Saha argued[54] that Eggert showed this ionization by applying Nernst's formula of "Reaction-isobar," K

$$K = \frac{p_M^{v_m} \, p_N^{v_n}}{p_A^{v_A} \, p_B^{v_B}} \dots$$

Here p denotes the partial pressures of the reacting substances M, N, etc.

Using this formula, Saha wrote the equation of the reaction isobar K, which is the equation of the gaseous equilibrium proceeds as given below:

$$\log K = \log \frac{p_M^{v_m} \, p_N^{v_n}}{p_A^{v_A} \, p_B^{v_B}} \dots = -\frac{U}{4.571T} + \frac{\sum v C_p}{R} \log T + \sum v C \qquad (5.1)$$

U denotes heat of dissociation, R is the gas constant, C_p is the specific heat at con-stant pressure, and C is the Nernst's chemical constant while the lines resulting from ionized helium are represented by the general series formula $v = 4N\left[1/m^2 - 1/n^2\right]$, and the best known of them that Saha identified in the visible range are the Rydberg line 4686 and the Pickering spectrum $v = N\left[1/2^2 - 1/(m+1/2)^2\right]$. The summation Σ in Equation 5.1 is extended over all reacting substances. Next, Saha went on to calculate U for considering the case of calcium denoted as Ca. We may regard the ionization of calcium atom as taking place according to the following scheme, familiar in physical chemistry—Ca $\leftrightarrow$ Ca$^+$ + e − U, where Ca is the normal atom of calcium (in the vapor state), Ca$^+$ is an atom that has lost one electron, and U is the quantity of energy liberated or absorbed in the process.[55]

The quantity considered, is 1g atom. To calculate the "Reaction-isobar" K, Saha assumed that P is the total pressure and a fraction x of the Ca atoms is ionized. This means the last term of equation can be written as using thermodynamics:

$$\sum v C_p = C_p\left(C_a^+\right) + \left(C_p\right)_e - C_p\left(C_a\right)$$

We can take $C_p\left(C_a\right) = C_p\left(C_a\right)$ and $(C_p)e = 5/2\ R$, the electron being supposed to behave like a monoatomic gas. Using Eggert's calculation for the chemical constant from the Sackur–Tetrode–Stern relation, Saha wrote the following equation:

$$C = \log\frac{(2\pi M)^{\frac{3}{2}}}{h^3} N^{-\frac{3}{2}} k^{\frac{5}{2}} = -1.6 + \frac{3}{2}\log M$$

Here, M is the molecular weight, and the pressure is expressed in atmospheres. Now C has the same value for Ca and Ca$^+$. For the electron M = 5.5 × 10^{-5}, which in turn, gives the calculated value of C = −7.99, Saha calculated ΣvC = −6.5. Substituting all the values obtained in Equation 5.1 using thermodynamics and atomic theory, Saha found the equation of the "Reaction-isobar" that is employed throughout for calculating the "electron-affinity" of the ionized atomas:[56]

$$\log K = \log\frac{x^2}{1-x^2} P = -\frac{U}{4.571T} + 2.5\log T - 6.5 \tag{5.2}$$

From this equation (Equation 5.2; later known as the *Saha equation*), one can easily deduce that when T is held constant, ionization increases as pressure decreases, implying that there is more ionization at the top of the chromospheres that stay at the bottom. For Equation 5.2, it is unclear whether Saha explored the condition that $x \neq 1$ and $x \neq 0$ i.e., $0 < x < 1$; otherwise, infinities occur. Using numbers for the various parameters gives a sense of the physical realization of the ionization numbers. For example, if photosphere temperature is taken as 7500K and pressure is 1 atmosphere, then for calcium $x = 0.34$, i.e., 34% of calcium atoms would be ionized.[57] At a different temperature at $T = 6000$K (further away from the center/core of the sun), using typical pressure value as 10^{-4} atmosphere, then $x = 0.95$, i.e., 95% of calcium atoms would be ionized, which means that at higher altitudes virtually all the calcium atoms are ionized, and this reasoning explains the "enhancement" of the flash spectrum that Lockyer had observed.[58]

Using a similar procedure, Saha explained ionization levels for other elements in the periodic table present on the solar atmosphere. Saha's paper "On Radiation Pressure and the Quantum Theory," submitted to *Astrophysical Physics Journal* in 1919, along with three other papers—"Ionization in the Solar Chromosphere," "Note on the Secondary Spectrum of Hydrogen," "Elements in the Sun"—submitted to *Philosophical Magazine* in 1920 from Calcutta, made a deep impact in the bourgeoning field of theoretical astrophysics by inspiring subsequent work undertaken by Alfred Fowler, Edward Milne, Cecilia Payne, and Henry Norris Russell.[59]

After communicating his papers to the *Philosophical Magazine*, Saha received a research leave—a Roychand Scholarship—from Calcutta University and traveled to London to work in Fowler's spectroscopy laboratory at the Imperial College (Figure 5.1).

While in London, on September 20, 1920, Saha applied to the Director of Special Enquiries and Reports, Board of Education, for the Chair of Physics post at Dacca University. Saha's letter, dated September 20, 1920, reads as follows:

> Sir, I beg to offer myself as a candidate for the Chair of Physics in the Dacca University. I was appointed University lecturer in Physics in the College of

Farewell Party to Prof. M. N. Saha at the University College of Science & Technology, Calcutta, on the eve of his departure to U.K. in 1920

Standing from left: H. Mitra, G. Dutt, D. D. Banerjee, S. K. Mitra, S. K. Acharya, A. C. Saha, A. N. Mukherjee, B. B. Ray
Sitting on chair from left: S. N. Bose, P. N. Ghosh, C. V. Raman, M. N. Saha, D. M. Bose, B. N. Chuckerbutty, J. C. Mukherjee

Figure 5.1 Saha in 1920 in Calcutta seated at the centre. To his left is Raman and Satyendranath Bose is seated to the extreme left (photo credit: *Science and Culture*).

Science, Calcutta in 1916 a post which I still intend to hold. Besides lecture work, I was in charge of the Heat laboratory of the College. I have previously come to England with the object of carrying out research work in Thermoionics and Thermal-radiation. I beg to enclose herewith an account of my academic career, and a list of my publications and some testimonials. As regards my works, reference may be made to Professor Dimarium of King's College, Professor Fowler of the Imperial College and Professor Lindemann of the University of Oxford.

> I have the honour to be, Sir
> Your obedient servant, Meghnad Saha[60]

Saha's tone in this letter is humble, considering he used phrases like "beg" twice and "your obedient servant." Though this kind of language was standard for letters of application in colonial India and may have been typed by a typist, it can be inferred from Saha's tone that, probably because of his lower caste origins, he felt the need to justify apologetically to the administrative and academic faculty of the upcoming Dacca University. Saha was going to be the first person from the *shudra* community to occupy a chair of any department, if Dacca University accepted him. He also was not famous enough to have a secretary, as he did later. During those days, most of the top positions in colleges and universities were monopolized by either the British or upper-caste Indians. Not over-interpreting the letter's deferential style, as it might have been endemic at the time, the incident shows what measures Saha had often to take in life to establish himself as a *bhadralok* scientist in his society.

While working at Fowler's laboratory, Saha collaborated with Fowler's spectroscopy group, and earned the trust of his British host. Before Saha left for Walther Nernst's laboratory in Berlin, he received a letter of recommendation from Fowler, addressed to the professor of Astrophysics and Fellow of the Royal Society, dated March 2, 1921 (Figure 5.2):

> Dr. M.N. Saha has been working at the Imperial College for the past few months and it is a pleasure to me to state that I have formed a very high opinion of his ability. He has a very extreme knowledge of mathematical physics and is well acquainted with the most recent developments in spectroscopy and the quantum theory. He has already made substantial contributions to science, showing distinct originality and independence of thought and I have little doubt that he will carry on his work with success. He is an industrious worker and we have found him personally very agreeable. He seems to have complete command of the English language and also reads German easily.[61]

This is high praise indeed for a student who came from a colonized country, struggling to establish himself as a scientist through his intellect. Fowler's letter negates many Orientalist stereotypes about India and Indian scientists not being able to produce new scientific knowledge. In fact, Saha himself, while commenting on

Figure 5.2 Meghnad Saha (photo credit: *Science and Culture*).

the Raman Effect and his own contributions to modern science, made the following remark echoing a reversal of Orientalist stereotypes:

> This is rather strange, because European writers are never tired of describing
> the Indians as given over completely to metaphysical speculations, and pos-
> sessed of little practical abilities. Here the roles are reversed—an Indian giv-
> ing the first practical effect to the theoretical speculation of European savants,
> which they themselves have been unable to verify.[62]

In this remark, Saha referred to Adolf Smekal, Kramers, and Heisenberg as
European savants. By European writers, he meant James Mill and his very popu-
lar text *A History of British India*, which is filled with such Orientalist stereotypes.
Not only was Saha well versed in the events that had led up to the discovery of the
Raman Effect, but he was also aware of the social and political underpinnings of
the importance of this discovery for India.

Still, relationships in science were not always hegemonical. As we see in the Saha–Fowler correspondence, there was a dialogue and accommodation between the scientific metropole and the colonial periphery, leading to the benefit of science. Another letter, dated March 17, 1921, from W.B. Hardy and J.H. Jeans (secretaries of the Royal Society of London), also revealed this correspondence by informing Saha, that after considering his application for a grant-in-aid, The Royal Society allocated fifty pounds for his research on thermal ionization of gases.[63]

In 1921 at the University of Berlin, Saha performed experiments on the electrical conductivity of heated caesium vapor. These experiments required a facility that housed a high-temperature physics lab. As such, an experimental facility was missing for Fowler's research group at Imperial College, so Saha's move to the physical chemistry laboratory of the University of Berlin to work with Nernst made sense. Every week he attended the colloquium at the University of Berlin. Therefore, he got to meet several well-known scientists like Max Planck, Albert Einstein, and Max von Laue. After getting an invitation from Arnold Sommerfeld, he visited Munich to give a colloquium. The experimental results Saha obtained at Berlin with Paul Gunther were subject to some scrutiny by Nernst, who considered the results rather uncertain and qualitative.

Due to a shortage of time, Saha had to leave Germany for India, as Ashutosh Mukherjee wanted him to take up the newly created position of the Khaira Professor in the Physics Department at Calcutta University in 1922. In a letter to Saha, on February 9, 1921, Mukherjee wrote:

My dear Meghnad,

Yesterday a cable was received at the University from Sir William Meyer, High Commissioner for India, stating that you had applied for a grant for 125 pounds and proposing that the University (of Calcutta) should finance you. We have sent a cable to the High Commissioner requesting him to pay the money to you at once on our behalf and we are repaying the money to him by a draft which will follow by next mail. I wish you had applied to your Alma Mater and not to the High Commissioner. We are in great financial crises here on account of the Non-Cooperation movement but you may (be) rest assured that so long as it is practicable your Alma Mater will not be slow to help you ... I have read with much interest your latest papers and I feel proud of the work accomplished by one of my fellow graduates. I trust you will not hesitate to serve your Alma Mater when you return.[64]

The University of Calcutta, Saha's alma-mater, was going through a deep financial crisis. Despite such an unstable financial and political scenario, Mukherjee[65] was keen on getting Saha back from Berlin and into the University, so that the latter could pursue scientific research for the progress of his nation.

Subsequently, as Saha did not get positive feedback from Nernst, he published his co-authored paper with Paul Gunther "On the Ionization of Gases by Heat" in *the Journal of the Department of Science* of Calcutta University. This paper was based on Saha's experimental work at Nernst's lab, and examined whether gases like Caesium could be made to conduct electricity by thermal ionization. Caesium was chosen, as it had the least ionization potential. As previous experiments on this problem were indecisive, Saha found the answer using the Richardson equation for thermoelectric emission $i = A\ T^{1/2}\ e^{-\varphi/k\tau}$. Here the symbol "i" was the emission current density, T the temperature, "φ" is the work function, and "k" is Boltzmann constant. The object to be found here was the work function.

The experimental results showed that conductivity depended on the Caesium vapor content under consideration. The conductivity of the vapor space became almost double when the temperature was raised from 1050 degrees Celsius to 1250 degrees Celsius. While experimental difficulties and errors had to be incorporated, Saha and Gunther concluded that the origins of conducting free electrons were because of the ionization of the atoms of these metals by using only heat.[66] This result also explained Saha's earlier researches found in his paper, "Elements in the Sun" in 1920, which suggested the absence of certain elements like Caesium from the solar spectrum.[67]

After the European sojourn

Saha returned to Calcutta University in 1921, and found that the institutional mechanics had changed. Ashutosh Mukherjee had set up the postgraduate department of the university, despite strong opposition from Lord Ronaldshay, the Governor of Bengal. While Mukherjee wanted Calcutta University to be a teaching and research university, the colonial government wanted it to remain a mere examining body. Consequently, Mukherjee faced a severe financial crunch and found it very difficult to hire research assistants and pay salaries to professors. Saha realized that it was virtually impossible for him to set up a research laboratory in the foreseeable future. Meanwhile, his college friend, Nil Ratan Dhar, a professor of chemistry, joined Allahabad University in 1919 and was actively engaged in research activities. Saha began looking for professional opportunities outside Calcutta, especially in Northern India, to create a school of physics. In 1923, he accepted an offer from the University of Allahabad arranged by Dhar. He stayed in Allahabad as professor and head of the physics department until 1938.

In Allahabad, Saha had to grapple with a higher teaching load and a lack of colleagues at the physics department. Though Calcutta had a more intellectually stimulating atmosphere, Saha did not get along with C.V. Raman, who disagreed with him on more than one occasion. While Saha might have felt marginalized in Indian society due to his humble background, he commanded respect from the international community of academicians, who knew him as a well-mannered middle-class intellectual of India. French litterateur Sylvain Levy, who came to India to meet Rabindranath Tagore, wrote to Saha on November 8, 1921 the following letter:

My dear professor,

I spent just this day at Calcutta going to Bolpur. Of course I shall come again here for a longer stay and I am sure to visit you that time. Now I send you through Rathindranath Tagore's kind officer, the fifteen Rupees M. Vidal had to take from your pocket at Calcutta for, after his unfortunate landing. I promised him to pay the amount as soon as possible;

I know he will write himself to you to thank again. I am, my dear Professor,

Your very obliged pupil, Sylvain Levy[68]

Levy's letter demonstrates the degree of respect held by Saha, especially as Levy went to the extent of calling himself a "very obliged pupil" of Saha at the end of his letter, notwithstanding the fact that he was never actually taught by Saha. Hence, we see that Saha was known to academics of the outside world as a distinguished scientist of India.

Social dimensions of science: Rivers and water

So far as application of science and technology in national life was concerned, Saha's driving force was humanism. As a scientist, he felt that he had a commitment towards his own nation, which was in a "backward" state. With the help of science, he wanted to modernize his nation in various ways so that the condition of the greater public would improve. Although he was basically a physicist, he could not afford to remain indifferent toward the devastation caused by natural calamities, rendering millions of people homeless.

Saha's approach toward rivers and water, along with methods proposed to reconstruct railway lines, is a characteristic that highlights the social dimensions of his science. For instance, in 1913, the Damodar River in Burdwan had a devastating flood, in which thousands died or were rendered homeless.[69] At the time, Saha was pursuing his M.Sc. at the University of Calcutta, and was deeply affected by this catastrophic event. He became a volunteer and worked towards helping the victims. In 1922, North Bengal was badly hit by floods, and barely any relief was provided by the colonial administration. Saha had just come back from Germany, after finishing his post-doctoral work.[70]

Along with P.C. Ray and Subhas Chandra Bose, Saha got involved in the flood relief programs and raised 23 lakhs[71] of rupees for the flood-affected victims. However, he was unhappy with the flood control measures available at that time, and felt the lack of a scientifically devised method for a permanent prevention of floods. Since his childhood, he has been fascinated by rivers, considering that he was from East Bengal, which was a land of many rivers. He always envisioned rivers as a grace of God, and a source of power for mankind. He wrote an article in the *Modern Review* entitled "The Great Flood in Northern Bengal," in which he remarked:

It has been said that Bengal is the "gift of the Ganges (and of the Brahmaputra)". But along with gifts, not infrequently, come curses, and it was the sad lot of the people of North Bengal to have a taste of these just a few weeks ago … To understand the nature of the havoc done, one must have a look at the railway lines in the district. Several lines are insufficiently provided with culverts, and oftentimes the culverts of the meter gauge lines have no corresponding culverts on the parallel broad gauge section. The floods have taken the same course as in 1918, and that this was due to the construction of the Sarah-Serajgunge railway. The railway [lines] were not breached and the railway authorities including Col. Cameron, who happened to pass by the spot a few days later, congratulated themselves on the strength of the lines built by them. But this joy in the railway camp was mingled with the wails of distressed villagers living in the affected area.[72]

As stated earlier, Saha always wanted to use science for the development and progress of his nation. At the same time, he sympathized with the people of Bengal suffering from natural calamities, like floods. He wanted to alleviate the distress and anguish of the people, and this desire motivated him to apply newer scientific techniques in order to better prevent and prepare for the possibilities of natural calamities as much as possible. His sympathy for the Bengalees reveals traces of subnationalism[73] in his mind, which was not in conflict with his nationalism.[74]

Saha's approach to alleviating natural disaster problems placed emphasis on designing proper culverts to improve drainage mechanisms for accommodating natural water flows after the construction of a canal. This was also a period when canal engineering, a nineteenth-century idea, had different modes of functioning compared to railway engineering, a twentieth-century idea. Saha was more inclined toward the canal engineers, and wanted to help them as an advisor as well as a financier. He realized that the British engineers had constructed railway lines for the administrative expediency of the colonial government, thereby completely disregarding the harm caused to the lower-class inhabitants of the villages close to the rivers and on their fertile lands. With the absence of culverts in the railway lines, there was no space for the water to drain out, which led to the accumulation of high levels of water.

Saha reasoned that railroad engineers trained in England had not considered the importance of unobstructed flow of water flooding the villages, and commented:

the havocs done by the present flood and the flood of 1871 are identical in all respects except as regards the destruction of crops. In 1871, the food was not held up by railways embankments. The water rose slowly and subsided rapidly enabling the *aman* crop to survive the deluge. But this year, the railway embankments held up the water so long, that the crops became a total loss. I think that if the railway embankment were provided with waterways, or if during the early stages of flood, the line were breached or cut open to let the water pass freely, the rice crop of the Rajshahi district, and the *ganja* crop of Naogaon could have been largely, if not wholly, recovered. Such a precedent of cutting the line in time for public good is not unknown, even in India … If the Railway engineers who constructed the Sara-Santahar and

Sara-Serajgunge were conversant with this episode, and cared to take their lesson from it, the people of Rajshahi and Pabna would have been spared much of the misery to which they have been subjected.[75]

Saha wrote several articles on floods and their prevention in the Indian journal *Modern Review*, in which he explained the river problems in India, and emphasized the importance of a hydraulic research laboratory for the study of rivers and floods in Bengal.[76] He wanted to create a research institute after the model of the Wasserbau Laboratory of Berlin-Charlottenburg, that would study rivers and formulate plans to simulate them in the laboratory. He believed that the laboratory should have close ties with a university having proper research facilities. He suggested that information about the periodic variation of all the rivers in Bengal, as well as the amount of silt brought by them, the distribution of water, and the precipitation data for each river basin be collected and integrated with the living conditions of people in rural areas before undertaking any engineering work. He was also aware of the devastating Damodar river floods, and suggested building storage basins in the Chotanagpur area at the points where the Damodar emerged from the hills.[77]

With the help of his student, Kamalesh Ray, Saha collected the historical records of the Damodar flooding. He was inspired by the dams used for flood control constructed in the United States by the Tennessee Valley Authority. Due to his own scientific knowledge, he did not hesitate to use the Wasserbau Laboratory of Berlin-Charlottenburg as a model for the welfare of the nation. He was always ready to accept the technical know-how prevalent in the West. He remarked:

> It is rather strange that in these days when Science is being applied to every walk of life for increasing human comfort, this problem of river control has never been scientifically studied in this country ... The idea of a River Physics Laboratory is not new. England has no laboratory of this kind ... Hence English engineers of the past generation have not been alive to the necessity of having such a research laboratory. But every other civilized country possesses a number of private and public laboratories attached to Technical High Schools or Universities ... Germany has been the pioneer, as in many other enterprises, in the development of these laboratories ... America comes to the opinion that the Physics of Rivers should be first studied in the lab before any great engineering problem is undertaken. My final suggestions are: (a) Establishment of a Hydraulic Research Lab for study of the problems of River Training, Flood Irrigation, Navigation, and Waterpower development in Bengal. This should be a purely Research Institute after the model of the Wasserbau Laboratory of Berlin-Charlottenburg or Vienna. The object should be the study of the physics of Great Rivers, preparation of plans in combination with department (b) and testing of the plan by means of lab models.[78]

Giving such suggestions, Saha outlined the importance of research laboratories as they functioned in the West, emphasizing the role of the physicist in such endeavors when he commented:

As the problems require expert knowledge of physics and mathematics, and demand much originality for their solutions, the laboratory should have a research atmosphere. It should be placed under a distinguished physicist who is also well up in mathematics. He should be provided with a good staff consisting of experts in allied lines and a good lab. Such a lab should be attached to the Universities, as Engineering Colleges in our country have not yet developed any research atmosphere. The initial expenses of a lab should not exceed Rs. 10 lakhs and the recurring expenditure 2 lakhs. (b) Department of Field Service: This will undertake a hydrographic survey of the rivers of Bengal, including relevant topics in Topography. Collection of Precipitation Data (such work is being done on a small scale by Prof. P.C. Mahalanobis in the Presidency College), and other geophysical factors likely to be of use in the preparation of great constructive projects. The department may be easily financed if my proposal of imposing a small thoroughfare tax on the passengers and trading parties using the E.I. Railways and E.B. Railway lines are accepted.[79]

Saha felt that the physicist's role in national life should not be underestimated. Building a river-physics laboratory would benefit the public, and would additionally decrease the number of unemployed physicists in India. A close collaboration of science and engineering would also be helpful for such a purpose, so that theoretical problems related to river research could be dealt with properly, along with field surveys and organized construction projects by people with technical expertise. This concern of Saha was motivated by his humanitarian outlook that made him willing to use science for the benefit of the nation in order to solve large problems—such as poverty, malnutrition, illiteracy, unemployment, and the "like evils," so that the nation could "build a happier, fuller and more prosperous life for the common man."[80]

Science and the nation

Science for the nation was one of the major driving forces for Saha. His initial engagement with his motivation came from the efforts he made toward flood control. At the same time, he took active interest in science as a professional occupation. While in Allahabad, in the state of United Provinces, Saha was slightly isolated, and unfortunately, the physics laboratory was not well equipped. Things changed in 1927 when he was elected Fellow of the Royal Society of London. His laboratory got a research grant of 5000 rupees per year, sanctioned by Sir William Morris, the Governor of United Provinces (U.P.), as a token of appreciation for Saha's international recognition. The financial issues of hiring research assistants improved, so that Saha had N.K. Sur, P.K. Kichlu, D.S. Kothari, B.N. Srivastava, G.R. Toshniwal, A.N. Tandon, S.C. Ray, Saradindu Basu, A.S. Mathur, S. Malurkar, and K. Majumdar as research associates in his laboratory. Research progress continued in theory and experiment with the help of the governmental grants. A new spectroscopic laboratory was set up by Saha during 1923–38. Thus, Allahabad became one of the leading centers of physics in North India.[81]

In 1931, the United Provinces Academy of Sciences was founded because of Saha's efforts. A year later Saha became President of the United Provinces Academy of Sciences. He played a key role in forming the National Institute of Sciences in 1934 and helped in the establishment of the Indian Science News Association (ISNA). With the help of ISNA, he founded the journal *Science and Culture* in 1935. In the opening issue, he highlighted the social functions of science and wrote the following editorial about electricity and its use for the public and for industries saying:

> In this article we wish to invite the attention of our readers to the much neglected problem of public supply of electricity which, now-a-days, is a main factor in economic progress and in which the public in this country has not so far taken any intelligent interest. It is clear that when the control of a commodity of such vital importance as power supply passes into the hands of a few big companies presided over by powerful bosses, there arises the danger of great misuse or privileges, for there is a risk of exploitation of the public for the advantage of a few ... The underlying idea is that electricity should be supplied at as cheap a rate as possible to the public for all purposes, and an easy and uninterrupted supply of it should be secured just like the supply of water to a big city ... electric supply for domestic use should be as cheap as in London, because it possesses all the advantages of London. But the actual figures show that the rates are eight times higher than in London. In a city like Allahabad the rates charged for domestic consumption should be the same as Portsmouth, but actually rates charges are two and half times higher.[82]

Saha was tormented by the fact that public supply of electricity was at the mercy of big companies, which "are plain and simply profiteers."[83] Electricity as a "public utility concern"[84] was of crucial importance to him, especially as seen in England and the United States where the public was ensured a cheap supply of electricity, and restrictions against profiteering and exploitation were in place. Analyzing the causes of India's backwardness, Saha wrote another editorial in *Science and Culture* in 1935 that said:

> We may now ask ourselves why the total production of electrical energy is so small (in India). The reason may be (i) that India has no adequate power resources; (ii) that the resources exist, but for some reasons have not been developed; (iii) that the existing laws relating to generation and consumption of electrical energy are such as to retard growth of the industry. The first alternative can be ruled out, as according to competent authorities there is plenty of power resources in India ... Neither Southern nor Western India has much coal. But this is compensated by their possession of magnificent hydro- electric power resources in the form of waterfalls and hill rivers whose water can be impounded in lakes at a high altitude; very little of this power has been developed.[85]

There were two main concerns that motivated Saha to start the journal. One concern was to emphasize the importance of the role of science and technology for national reconstruction. As expected, *Science and Culture* published contributions made by political thinkers, scientists, and the general intelligentsia, articulating a vision of a modern industrial nation through scientific and systematic planning. In June 1935, Saha remarked:

> The call that brings "Science and Culture" into existence is truly the call of the times. For it is obvious to every thinking man that India is now passing through a critical stage in her history, when over the cultural foundations of her ancient and variegated civilization, structures of a modern design are being built. It is necessary that at such a juncture the possible effects of the increasing application of discoveries in science to our national and social life, should receive very careful attention; if the present is the child of the past, it may with equal emphasis be asserted that the future will be the child of the present.[86]

The main emphasis of the articles published included measures to be taken for eradicating India's poverty and unemployment through modern industrialization. The Soviet model of planning and state-directed growth, along with the American model of agricultural research and development, served as the source of inspiration. Saha used the journal frequently, usually in unsigned editorials, to expound his views.

Apart from *Science and Culture*, the other contemporary journals in the 1930s included the *Kesari*, which published articles in Marathi language about nationalist politics; the *Dawn*, which published works on philosophy and religion; and *Transactions of the Bose Institute*, which focused on physics and biology. Most of them were either too technical or too political for a general audience to appreciate, whereas Saha's journal included articles written in popular language by experts on various branches of science and government policies, covering technical matters like rural reconstruction, transportation, power development, river valley development, and so on. There was also space reserved for correspondence among scientists about the validity of scientific theories and their further ramifications, and exchange of short research notes on scientific and cultural issues.

The editors of the first thirteen issues of the journal published in 1935 included Saha himself from the University of Allahabad, B.B. Ray from Calcutta University, who was also one of the founding secretaries of Indian Science News Association, and Jnan Chandra Ghosh from Dacca University. Interestingly enough, Ghosh, a classmate of Satyendranath Bose and Saha in Presidency College, circa 1910, was the founding chairman of the department of chemistry at Dacca University in 1921.

The first volume of *Science and Culture*, which came out in June 1935, contained four articles by Saha. The first article, "Ultimate Constituents of Matter," discussed recent advances in nuclear physics, including the atomic model of Niels Bohr, in addition to developments in radioactivity. Then in "The March

Towards Absolute Zero," Saha pondered over the post-doctoral work he had done in Germany in Walther Nernst's lab, on thermodynamics and statistical mechanics. Next, "The Existence of Free Magnetic Poles" dealt with theoretical speculations of British physicist Paul Dirac (of Dirac equation fame) regarding the possibility of isolating a magnetic pole. Furthermore, "Spectra of Comets" featured Saha's own research in astrophysics in 1921, explaining the deviations in the sun's spectrum by observing the spectra of comets. These articles were up to date with scientific discoveries in physics in Europe and India. With his journal, Saha attempted to guide the minds of the common people of his country toward a more science-oriented comprehension.

Robert Anderson argues that Saha thought *Science and Culture* would play the role that the journal *Nature* did in Britain, or the journal *Science* did in the United States. Anderson points out that Saha's journal would interpret science in non-technical language and advocate a planned application of science on India's unique problems.[87] Whether Saha was aware of the difference in literacy levels in India, England, and America, however, is relatively unknown. Although Saha's intention was to convey science to the public, sometimes the articles published in the journal were not non-technical.

Moreover, the journal proclaimed that the "future belongs to those who know how to use machines as slaves and not ask human and animal muscles to bear the strain which machine can bear."[88] The journal wanted to promote close links between science and industry. It also published articles by Nehru and Subhas Chandra Bose, who used it as a forum to express their vision of a modern India. Regarding this issue, Nehru wrote:

> We have vast problems to face and solve. They will not be solved by politicians alone, for they may not have the vision or the expert knowledge; they will also not be solved by scientists alone. For they will not have the power to do so, or the larger outlook which takes everything into its ken. They can and will be solved by the co-operation of the two for a well-defined and definite social objective.[89]

In a similar vein, Subhas Chandra Bose wrote:

> The appearance of Science & Culture is to be warmly welcomed not only by those who are interested in the abstract sciences but also by those who are concerned with nation building in practice. Whatever might have been the views of older "Nation-builders", we younger folks approach the task in a thoroughly scientific spirit and we desire to be armed with all the knowledge which modern science and Indian culture can afford us ... I shall now ask our scientists to take up these problems one by one and give satisfactory answers. Without the co-operation of science, no nation building is possible.[90]

Moreover, Saha's second concern in establishing *Science and Culture* was to compete with the Bangalore (South India)-based *Current Science*, another science journal introduced by Raman in 1932. The conflict between Raman and Saha had its origins in the Indian Association for the Cultivation of Science (IACS) in 1917, when Raman became the first Palit Professor of Physics (the most honored position in the Calcutta University). In the IACS, Raman attempted to confine the membership of the institute to only South Indians. This created problems for the institute and other senior members who were mostly Bengalis. Later in 1931, the confrontation between Saha and Raman came to the public eye. Saha was very much against parochialism, as he believed it could act as a stumbling block to scientific progress.

Saha's scientific attitude was tempered by his concern for human welfare, and he always felt for the proletariats of society, the subaltern sections of the population. He was conscious of the "humanizing influence of science," as he remarked:

> If there be a rational programme of production, and a programme of judicious and equitable distribution, nobody should suffer from hunger, privation and can even afford to have much better amenities of life. But for this purpose, rivalry among nations and communities should give way to co-operative construction, and the politician should hand over many of his functions to an international board of trained scientific industrialists, economists and eugenicists, who will think in terms of the whole world as a unit, and devise means by which more necessities of life can be got out of the earth.[91]

Saha felt the need of modernization with the help of scientific and industrial progress in a harmonious way through cooperation between nations. Only through a proper balance between politics and science would a perfect nation-state, as visualized by Saha, emerge. This vision made his worldview similar to Nehru's views, though Saha's approach espoused nation-building by ameliorating the basic problems of the lowest rungs of Indian society; somewhat dissimilar to Gandhi's utopian views (as we will see later) which were ambivalent towards science.

It may be relevant here to point out that in 1922, Saha was invited by Subhas Chandra Bose (Subhas) to the Bengal Youth Conference to deliver a speech on the economic reconstruction of India after decolonization. In his speech, Saha advocated the need for scientific development of large-scale basic industries. In 1938, after the Haripura session of the INC, he became a leading member of the National Planning Committee set up by Subhas, who was the president of INC at that time. This committee was the precursor of the present Planning Commission of the government of India that now formulates economic plans for every five years. In the 1935 October issue of *Science and Culture*, however, Subhas wrote an article entitled "Some Problems of Nation Building," in which he laid stress on the importance of a five-year planning system for a balanced economic development. Both Saha and Subhas Bose admired the system of a planned national economy prevalent in the Soviet Union under socialism. They believed that India could move forward with the help of science and technology used for industrialization in a well-planned manner.

On the margins of the nation

The way Saha addressed the issue of national reconstruction was more pragmatic than that of Gandhi and Nehru. As a scientist, Saha had the ability to assess the merits and demerits of the possible ways of reconstruction. As he possessed a very strong and self-confident personality, there was often a difference of opinion between him on the one hand and Gandhi and Nehru on the other.

Gandhi gave a critique of modernity in the West, and attributed India's poverty and backwardness to industrialization. He remarked "I am not against machinery as such, but I am totally opposed to it when it masters us."[92] To him, large-scale industrialization was nothing but a boundless increase of greed. At a discussion with his educationist colleague G. Ramachandran, Gandhi espoused a view which was against any fixation on machinery, and thus remarked:

> The supreme consideration is man. The machine should not tend to make atrophied the limbs of man. For instance, I would make intelligent exceptions. Take the case of the Singer Sewing Machine. It is one of the few useful things ever invented, and there is a romance about the device itself. Singer saw his wife labouring over the tedious process of sewing and seaming with her own hands, and simply out of her love for her, he devised the sewing machine, in order to save her from unnecessary labour. He, however, saved not only her labour but also the labour of everyone who could purchase a sewing machine.[93]

In effect, Gandhi articulated a utopian vision of independent India, by jettisoning modern scientific technology and reverting to the spinning wheel (*charkha*) and *khadi* (hand-woven cloth). With the consolidation of Indian nationalism in the 1930s, and national independence not very far away, Gandhi was still pondering over alternatives to the industrial approach. These alternatives (beyond the scope of this monograph) were not necessarily in keeping with efficient and more scientific approaches of the time, but those that espoused indigenous and self-sufficient ways as opposed to the juggernaut of industrial imperialism.

Gandhi contradicted himself on a few occasions (per Saha), such as when he praised the sewing machine as a labor-saving device, while at the same time he wanted his countrymen to take recourse to the spinning wheel and wear *khadi* instead of clothes produced in factories. In the ultimate analysis, we find his leanings were towards agriculture and cottage industries, not heavy industries with all their concomitants. He argued that:

> the spinning wheel and bullock cart should be protected so long as the state cannot provide for the victims of unemployment; on the other hand, there

should be unremitting efforts to adopt the modern technic to all the needs of industrial and economic life, and old antiquated methods should be discarded without a sigh or tear when the proper insurance against unemployment has been made. It is to preach this middle path that "Science and Culture" makes its debut before the public. Its object is dissemination of scientific knowledge amongst the public and advocacy of its application to all walks of life as far as practicable ... the view will always keep in mind that science is important as long as it conduces to the development of culture and serves the cause of human progress.[94]

According to Saha, progress in science and technology was essential as it helped a culture to evolve, and especially enable national reconstruction. Saha offered the following remark about Gandhi's (also known as the Mahatma) vision:

To us, scientists, it appears that the Mahatma's system lacks progressive vision i.e. it does not say how villages are to be linked to the cities, and how the industries which are indispensable for the Nation's life and for the body-politic are ever to be managed by Indians for the benefit of the Indian population. Apart from adopting a policy of *laissez faire* to these urgent problems, his whole attitude towards the machine and the modern city-civilization is one of defeatism. He looks at its evils, but does not try to understand its mechanism of work and he starts with the inner conviction that the machine- civilization must be intrinsically evil ... It would be a happy day for India if Mahatma can overcome his attitude of defeatism towards the Machine, devote a little time to the mastery of the technique of modern civilization, and then make up his mind. Otherwise we feel that by diverting the attention of the Nation from the only path which holds out prospects of relief against the present problems of poverty, unemployment and defenselessness, he will be committing what we may describe by the oft-quoted phrase as a 'Himalayan Blunder.'[95]

Saha's disagreements with Gandhi were also on the best trajectory to achieve industrialization and its outcomes. While Gandhi argued that the Industrial Revolution in Europe caused great social dislocation and political unrest, Saha countered this argument by using the example of Japan, which had been industrialized without having to borrow foreign technicians or foreign capital. He said:

Whereas the story of Japan's rise as a great power is sufficiently well-known, the secret of her success in her experiments with modern civilization, the extent of her political and economic problems, and the vastness of her ambition with their bearing on India are not properly realized or appreciated ... India, as well as China, burdened with her great past looks back with longing, lingering eyes to the vanished village economy, to the cult of the spinning wheel and the bullock-cart and primitive Agriculture and primitive Industry as the panaceas for all her ills. Japan did not commit that mistake but threw herself, with all her youthful energy, into developing her natural resources in minerals and power, overhauling her industrial machinery, and organizing an efficient system of national education…And she has done all this, without having to import foreign capitalists or foreign technicians, or without having to plunge into a bloody civil war.[96]

Their disagreements were also on Gandhi's political philosophy, which advocated non- violence (*Ahimsa*). Professor B.P. Agharkar's monograph, "If the War Comes," written in 1939, was in tune with Saha's critique of Gandhi. In a foreword for Agharkar's book, Saha commented:

The largest and the most powerful party seems to be of the idea that if the menace of foreign aggression ever materializes, India should not attempt to meet violence with violence, but will be better advised to practice *Satyagraha* (non-violence) and thus melt the heart of the aggressors. This is in consonance with the political philosophy of Mahatma Gandhi. In the book before us, Prof. Agharkar very ably refutes all these fallacies, which are … that non-violence (*Satyagraha*) can never be an effective weapon against foreign invasion and in internal quarrels.[97]

This remark also reflected Saha's views which tended to agree with other Indian nationalist leaders, such as Subhas Chandra Bose, who believed in armed resistance to bring an end to the colonial rule. As Gandhi cited Buddha to defend his theory of non-violence, Saha criticized him by saying that:

Perhaps for extolling the cult of non-violence, no great founder of religion is more misrepresented than Lord Buddha. On many occasions, Gandhi has preached that *Ahimsa* should be practiced and has cited Buddha in defense of his arguments. But let us see that Buddha himself thought about these matters … Buddha said to Simha (a general of the great Lichchhavi Confederacy) he who deserves punishment must be punished … for whosoever must be punished for the crimes he has committed, suffers his injury not through the ill will of the judge but on account of his evil doing. The Tathagata does not teach that those who go to war, in a righteous cause, after having exhausted all means to preserve the peace are blameworthy. Buddha completely justified wars undertaken in self-defense, and for the punishment of the wicked.[98]

Conversely, Saha and Gandhi also had some points of agreement, as both agreed on the necessity of a common script for India based on Sanskrit. A language for the nation based on the indigenous Sanskritic languages was the general goal of the Indian nationalists in the early twentieth century. However, here too, Saha had some difference of opinion with Gandhi on the specific details, which become apparent when Saha said:

Gandhiji's suggestions betray only an eagerness for a practical solution, but they are far from being helpful. "Hindi and Urdu" says Gandhiji, "will continue to flourish. Hindi will be mostly confined to Hindus and Urdu to Muslims ... Hindustani of the Congress conception has yet to be crystallized into shape ... For the purpose of crystallizing Hindustani, Hindi and Urdu maybe regarded as feeders ... The Hindustani will have many synonyms to supply the various requirements of a growing nation with rich provincial languages. Hindustani spoken to a Bengalee or South Indian audience will naturally have a large stock of words of Sanskrit origin. The same speech delivered in the Punjab will have a large admixture of words of Arabic or Persian origin. Similar will be case with audiences composed predominantly of Muslims who cannot understand many words of Sanskrit origin." Any thoughtful reader will see that Gandhiji has been spinning around a circle, without being able to suggest any progressive step at all. Do not Hindi and Urdu serve in Northern India the same purpose today which is sought to be served by Gandhiji's brand of Hindustani? It will be apparent to all clear thinking people that this sort of statement confuses the issues. Why not call these variants plainly as Hindi or Urdu? Gandhiji might have more frankly asked every Indian to learn both Hindi and Urdu and accept them both as national languages, as he has done in the case of scripts.

Really, if such different variants of Hindustani are to be used in different provinces and for different audiences, how can a common language crystallize into shape?[99]

By a common language Saha meant one that could be accepted by a pan-Indian audience, and which was free of local variations. For example, an acceptable language would be one that even the leaders of Congress could use while addressing a nationwide audience.[100] In this regard, Saha agreed with Nehru's analysis of the common language question. Nehru espoused a view that agreed to make a common language for India, while having the provincial languages remain dominant in the respective provinces. For example, with respect to Bengal, where Bengali is the dominant language, Nehru and Saha agreed that it did not make any sense to have a nationwide common language infringe on the domain of Bengali. Though points of agreement were found to exist between Saha and Nehru, they nonetheless differed on issues surrounding science and politics.

About Jawaharlal Nehru, it may be said that although he disagreed with Gandhi at times, he was undoubtedly the most trusted disciple of the Mahatma. Unlike Saha, Nehru used Gandhi's popularity in the Indian nationalist movement to strengthen his own position in the INC. Therefore, Nehru's occasional disagreement with Gandhi may be viewed through the lens of a give-and-take relationship. Saha, on the other hand, did not have any political ambition, as he was a scientist unfamiliar with the craft of Machiavellian politics. Saha never became a member of the INC, and his relationship with Gandhi developed through the NPC set up by Subhas. Nehru's view of Gandhi's science sheds light on this aspect, as he remarked in a letter written to Aldous Huxley in 1933:

Indeed, in a political sense, I have myself been a "Gandhi-ite" for all these years. His attitude to science is far from being hostile. He welcomes it and takes advantage of it in a variety of ways, and often people, willfully misunderstands him, accuse him of inconsistency because he does so, as for instance, when he submits to an operation, or rides a motor car, or uses a printing machine, or telegraphs or telephones. None the less, it is perfectly true that, being fundamentally religious, his tendency is to seek for the truth inside himself, rather than externally by the methods of science. Large-scale machine production he does not like but he has never suggested, to my knowledge, that it should be scrapped. He wants, as far as possible, to decentralize industry, and for this purpose he wants to take fullest advantage of the scientific method. It may not be a correct attitude; its logic maybe faulty. But it does not mean a negation of science or a destruction of machine industry. But personally I do not agree with it.[101]

Nehru was responding to a letter from Huxley that characterized the INC as anti-scientific.[102] It seems Nehru's refutation was not very convincing, and he vacillated between two opposing viewpoints. This response shows the ambiguities in Nehru's position while trying to remain an ally of Gandhi, and at the same time distancing himself in subtle ways from Gandhi's extreme positions. When Nehru reacted to Saha's views on science, he could freely critique them, since he had no obligation to support the latter. As Saha was teaching at the University of Allahabad and living in Allahabad, which was also the location of Nehru's residence, both of them interacted frequently because of this geographical proximity.

Also an important year in the history of physics, 1939 was when fission was discovered by Lise Meitner and Otto Hahn. When Saha came to know about it, he introduced nuclear physics as a special subject in the physics syllabus of M.Sc. at Calcutta University. In this year, he was also appointed as the Palit Professor (a position previously held by C.V. Raman and D.M. Bose) at Calcutta University, which made him return to his alma mater. This was also the time when he became a member of the NPC, with Nehru as its chairperson. As a result, Nehru had some obligation to respond to the queries of council members. By doing so, he was trying to pacify Gandhi, while also agreeing with Saha on many basic issues regarding science, thereby making it seem he was anti-Gandhian. Nehru's personality was characterized by some inherent contradictions akin to the nature of his *guru*, the Mahatma. While commenting on the eighth anniversary of Saha's *Science and Culture*, Nehru remarked:

The very name of this periodical signifies the two things which, more than anything else, India, like all progressive and civilized nations, must possess … Science is the very basis and texture of life today and without it we perish, or, what is even worse, slide back to barbarism … Yet, though we swear by science and accept it advantageously for many purposes, still the habit of unscientific approach remains. Vested interests, superstitions and out-of-date customs prevent the full application of the scientific and rational method.

Indeed it is in these times of war and crisis that the rational message of science is all the more necessary. So more power to *"Science and Culture."*[103]

The core of Nehru's faith in Western rational ideas consisted of his belief that growth was the most crucial aspect of social life. By growth, he alluded mostly to large-scale economic and industrial development using purposive planning. He was found to be more ambivalent on the measures of progress when he remarked:

You can define progress in a multitude of ways, but today in India, at any rate, we can think of it only in one precise way. Ours is an urgent way, how we can deal with urgent problems in so far as they affect hundreds of millions of our people. It is they who count and nobody else counts in the ultimate analysis. And if anything, that we do, does not directly or indirectly affect their lives and hopes – and what they live for and what they hope for – then, for the moment that is not progress, although ultimately in the long run it may lead to progress.[104]

Industrialization was also a measure of progress for Nehru but in a distinct way, which was more evident when he said:

Industrialization is not just putting money – a penny in the slot machine and out comes the factory. It requires trained human personnel to do it; it requires a mental approach ... industrialization affects the world because science came into the picture.[105]

According to Nehru, both scientific and industrial development guaranteed the prosperity and superiority of a nation or a civilization. According to this view, the rational West represented growth, while the Orient (India and China) remained stagnant. Having made this choice, Nehru located a rational past in India. To him, a rational India from the past was the "true" India that had to be retained, while "the dust and dirt of the ages" that have covered her up and hid her inner beauty and significance stunted her growth.[106] Nehru argued that growth and wealth resulted from greater control over, and understanding of, the processes of Nature that scientific knowledge made possible. Although greater knowledge did not signify greater wisdom, aimless pursuit of knowledge or science could lead to cataclysm.

On the political front, the 1930s was the era of nationalist consolidation with the Salt Satyagraha and the initiation of the Civil Disobedience Movement. Nehru had just returned in 1935 from his visit to Moscow, and Soviet Russia became the source of inspiration at this juncture for the left-wing nationalists in the Indian National Congress.

Socialism came to mean a culmination of democracy and embodiment of social justice, and began to be accepted as the goal of free India. Nehru expressed his fascination for Fabianism and science as the trajectory that the nation should follow.[107]

By Fabianism, he meant the gradual route to economic reconstruction through planning and industrialization, as opposed to revolutionary means.

Nehru and Saha had co-operated in some areas of common interest from 1935. Similarly, their disagreements also came to the surface of their relationship at this time. When Subhas Chandra Bose became the President of the INC in 1938, the All India National Planning Committee (NPC) was set up by Bose, with Nehru as its chairman. In Rabindranath Tagore's letter to Nehru on November 19, 1938, on how the former viewed the NPC highlights, Tagore remarked:

> The other day I have had a long and interesting discussion with Dr. Meghnad Saha about Scientific Planning for Indian Industry; I am convinced about its importance and as you have consented to act as the President of the Committee formed by Subhas Bose for the guidance of the Congress, I would like to know your views on the matter.[108]

The NPC was inaugurated by Subhas Chandra Bose on December 17, 1938. The Committee appointed twenty-nine[109]subcommittees that framed resolutions for the economic and social regeneration of India, with Saha as the chairman of the sub-committee for Power and Fuel and Technical Education.[110]

Saha's scientism

In 1938, before returning to Calcutta, Saha, in collaboration with The National Academy of Sciences in Allahabad, organized a symposium on Power Supply. Having thanked Nehru, Saha remarked:

> It was in the fitness of things that Pandit Jawaharlal Nehru has agreed to pre-side over this annual general gathering of scientists in India. His position in the country can be described by a phrase which Americans use with respect to Abraham Lincoln: 'First in war, first in peace', and next to Mahatma Gandhi, he occupies the first place in the hearts of his three hundred and fifty million countrymen. The time has now come for him to give a lead in peace time work of reconstruction and consolidation of the country.[111]

Despite having disagreements with Nehru (as we shall see in the forthcoming narra-tive), Saha held him in high regard, comparing him with Abraham Lincoln. Motivating Nehru to give leadership to the country was another concern for Saha, especially for the reconstruction and consolidation of India. When the NPC started working in 1938, their difference of opinions emerged on the surface of their relationship.

An interesting exchange of correspondence reveals the nature of the Saha–Nehru dialogue in the context of the NPC. Writing to Nehru on January 19, 1939, Saha remarked:

> I am sorry to hear that you are not quite satisfied with the work of the meet-ing which was held at Bombay. As you have seen from the papers, the mere

proposals have produced a certain amount of apprehension in the British capitalistic quarters. This confirms us in our belief that the proceedings so far conducted have been on the right line ... I found that the committee has entirely ignored the technical and legislative side ... As I remarked on the occasion of the meeting of the Industries Ministries at Delhi, if the committee is to do really useful work, the most practical course would be that the majority of the members should remain at one place continuously for three months and give their undivided attention to the subject. The members of the committee (NPC) represent various ideologies and interests and it is but natural that during the first meetings, the views would be divergent. But if we go on studying the subject and discussing continuously then it is not impossible for a group of earnest-minded men striving at a common ideal to arrive at a common and useful formula. I would suggest therefore that the Planning Committee should meet for three months at Poona from 15th April to the beginning of June and all should reside in one house to be lent by the Bombay Government. I am making this proposal because, after all, physical comfort is necessary for serious mental work and according to meteorologists, Poona has got the best climate in India during this season. Further, it is one of the seats of the Bombay Government and is very close to Bombay. Thirdly, those of us, who are university men, will find it possible to give their undivided attention to the problem only during the summer vacation, when we are free from our routine work. I shall be glad if you can kindly agree to this proposal and write to the Secretary of the Planning Committee to take the necessary action.[112]

Nehru responded on January 25, 1939:

I am afraid I cannot issue any directions to the secretary of the Committee about our future work ... So far as I am considered it is quite impossible for me to give three months or even three weeks to this work. But of course the Committee can meet without me. I do not personally think at this stage it will be worthwhile for us to meet for a long period. There must be an intervening stage of sub-committees before the full Committee can function satisfactorily.[113]

Nehru was also aware of Saha's background as a scientist, and his own lack of expertise on scientific topics. This might be the reason for Nehru's statement concerning the Committee meeting without his presence.

Saha then replied on May 9, 1939:

I am glad to inform you that Mr. K.D. Guha has taken up his work seriously. I have given him a room in my laboratory and he is coming everyday and working. I have asked him to prepare the agenda for the next meeting, He will actually draft the resolutions which have to be moved to Bombay and also find out the personages who will speak in support of these resolutions and also those who will support them. I think after the agenda is framed and you approve of it, we have to write to these personages, so that they may be ready with their speeches. The idea is that each resolution will deal with a sub-committee which will go deeper into the thing. I have also asked several persons like Prof. D. Ghosh of Sydenham College, Bombay, who is now spending his vacation at Calcutta, to join Mr. Guha and help him in framing the agenda.[114]

Saha wanted to engage with Nehru as much as possible, and wanted to make sure they were not short on man power, especially when it came to drafting resolutions.

On May 13, 1939, Nehru wrote:

Our first steps will be to consider some basic matters of policy. Having determined this, we shall then proceed to consider the subjects in some detail and appoint sub-committees. Unless we are clear about our basic lines of policy it is difficult to go ahead or frame resolutions but ultimately they will depend on what the general policy agreed upon is and what our general objective is. I do not think, therefore, that we can at this stage have formal resolutions ready … My own idea of the Planning Committee's meetings is that we should proceed in a business like way with as little speechifying as possible. We have had a good deal of speech making at the last meeting.[115]

The above passage shows Nehru's reluctance to get into a discourse with Saha. His preconceived notions— "proceed in a business like way"—about how to progess did not include talking to a scientist, whereby Nehru might be getting into a discourse about the nuances of science with Saha.

Saha replied on April 12, 1940:

Neither the presidents nor the secretaries of many of the committees are members of the N.P.C. and in their absence it will be very difficult for us to consider the recommendations or the reports. I would therefore suggest that the presidents and secretaries of these sub-committees may be co-opted to the NPC. If that is not practicable, you may summon the presidents and secretaries to explain their recommendations before the NPC and reply to the criticisms which may be directed against their report. Unless this is done, I think it will be very difficult to make any progress with the reports of the sub-committees.[116]

Nehru responded on April 19, 1940:

As for your suggestion that Presidents and Secretaries of sub-Committees be invited to the NPC, I think this would be desirable … but I think that only those who have submitted full reports might be asked to come at this stage … otherwise there will be too much of a crowd.[117]

This correspondence between Saha and Nehru in the formative stages of the NPC represent some of the disagreements between them regarding the best way to handle the agenda of national reconstruction. It can be inferred from this correspondence that despite Saha's best efforts, Nehru was not always willing to accept Saha's proposals. In one of the letters (May 13), Nehru used the word "speechifying," which appears to be a deviation from the norms of official correspondence. Notwithstanding their differences, both Saha and Nehru employed a similar approach to the question of national reconstruction.

Both were nationalists, and both were inspired by the Soviet model of "planned national development." Saha's *Weltanschauung* regarded science, with its humanizing influence and industrialization, as an essential factor for the creation of a modern nation-state. What impressed Saha the most was the role of scientists in the Soviet Union in the development of the State. Saha visited the Soviet Union for the first time during the summer of 1945 as an Indian delegate for the 220th Anniversary of the Russian Academy of Sciences.[118] He quoted Gleb Maximilianovich Krzhizhanovsky, a member of the Russian Academy of Sciences, in his critique of Gandhi in the following passage:

> But what about the spinning wheel and bullock-cart economies of Gandhi? Clearly when he and his party get power and put these economies into practice, India would never walk out of medievalism, which would be far worse than British Imperialism, as far as the lot of common man is concerned. The spinning wheel uses human labor but you must be knowing that a man working for eight hours can produce work in the day which can be had only for 20 kopeks (3 pice in Indian money). We in Russia allow men to use either electrical energy or steam energy and every Russian worker does the work of seven or eight men. That is how we have increased our productive power and met Germany on almost equal terms.[119]

Saha agreed with Krzhizhanovsky's remarks about Gandhi and commented:

> Myself and many of my brother scientists have as little regard for Gandhi's economical and social theories, as you have for Tolstoy's and we have been putting forth the same arguments. You see that we were able to persuade the Congress High Command to set up a National Planning Committee with Gandhi's second in command, Jawaharlal Nehru, as Chairman. He enlisted the co-operation of India's best scientists, economists, and industrialists and our line of thought has evidently made some progress.[120]

Furthermore, Krzhizhanovsky asked Saha about Nehru and questioned:

> I have heard of Jawaharlal Nehru, and his National Planning Committee. Nehru is second in command to Gandhi, but does he also share Gandhi's economic ideas?[121]

As it was a difficult question to answer, Saha tried to bypass the question while keeping his response as ambiguous as possible by stating:

> What impressed me about Nehru and was a decisive factor in my desire to serve the National Planning Committee was a little remark in one of his writings. Talking of 'Liberals,' a class of politicians in our country, who want to pursue a middle path, and thus satisfy every party like your Mensheviks and Kerensky, Nehru said if you ask a Liberal whether the earth is round or flat, he will say neither the one nor the other, but after hesitation will say it is probably elliptical.[122]

This exchange of dialogues between Krzhizhanovsky and Saha reiterates the differences between Saha and Nehru in their respective outlooks.

Saha wanted to dissociate science and industry from political conflicts of the era, as reflected in *Science and Culture*, while allowing both science and industry to coexist independently. This aspiration is seen in *Science and Culture*'s articles that revealed admiration for Soviet, American, and British models of industrialization, while disregarding the ideological tensions between these industries. It may be said that Nehru was a very crafty political person, and knew about his iconic image within the Congress. He was, presumably, apprehensive of the possibility that the key decisions taken by scientists like Saha might jeopardize his own vested interests in the Indian political scenario.[123]

As the chairman of the subcommittee on Power and Fuel in the NPC, Saha wanted to make a comparative study of the average per capita income of the different countries of the world, including India. While doing so, he found it to be a difficult and an "almost hopeless task," so he decided to make a comparative analysis of energy production in different countries to assess the economic conditions prevalent in those countries.[124] While in the NPC, he wanted to induct reputed scientists of India as members of various subcommittees for their expertise.

Through the establishment of a scientific body like the Academy of Sciences, Saha wanted more involvement of scientists in national matters, and the use of science for nation-building. Thus, he commented:

> [The] ultimate purpose of the Academy ought to be; to recognize the scientific workers and to persuade them to take more interests in the scientific matters of national interest; and also to exercise a healthy influence on Government in its administrative policy regarding scientific matters.[125]

Nehru did not look favorably upon the idea of a scientist taking the lead in national politics and science related policies, as he was afraid of the fact that his ignorance in matters concerning science would be exposed while arguing with a fellow scientist, especially someone of Saha's stature. This tendency became a problem later, when Saha became a member of the Indian parliament in independent India, causing personalities to clash. However, the period after Indian independence is beyond the scope of this monograph.

Focusing on the case of Meghnad Saha, this chapter has explored how a lower-caste individual moved up the social ladder, and became a *bhadralok* through his pursuit of science. Though *bhadralok* was a category typically ascribed only to higher-caste intelligentsia, Saha's life trajectory shows a wide spectrum of people in this new group, who were also from lower social strata. Through Saha's lens, we also come to know how modern science managed to establish its roots and develop original directions of research in colonial India in the early twentieth century.

Saha's nationalism encompassed his close association with the Bengal revolutionaries, whose rationale was to use armed resistance against the British rulers for decolonization, as opposed to the Gandhian principle of non-violence. While not aiming to speak for Saha, this chapter has attempted to represent historical contingencies in a way that is conscious of the difficulties of science in a double power differential. Saha's science represents a specific brand of Indian modernity combined with a nationalist sentiment. He achieved the most significant scientific success in the revolutionary field of quantum physics, which included the Saha Ionization equation in 1921, thereby opening a new vista in physics. It was a major milestone in the route to quantum mechanics.

The social dimensions of Saha's science are to be traced in his works on rivers and water. His contribution in formulating engineering devices, like redesigning railway lines to avoid floods, indicates the humanistic dimensions of his science. His involvement with the Bengal revolutionaries reveals another aspect of Indian nationalism. We find that not only political participants, but also nonpolitical sections of the population had a strong nationalist sentiment. Saha's founding of the journal *Science and Culture* in 1935 emphasizes his great concern for nation-building with the help of science. From its inception, the journal sought to spread messages of science amongst people and make them capable of rational thinking.

The process of historical investigation, for me, is not restricted to a narrow engagement with elite figures like Mahatma Gandhi or Jawaharlal Nehru, but is an endeavor to situate it in an extensive horizon, involving a participant who was born in a marginalized community and therefore overlooked in traditional historical narratives of science in South Asia. The social and intellectual milieu in which Saha was raised significantly influenced his career. It is perhaps appropriate, then, to regard his most significant scientific successes—including the Saha equation—as important milestones in the unfolding of quantum mechanics.[126] And it is surely significant that he used science to promote a spirit of nationalism in colonial India, as well as a distinct national identity as a *bhadralok* in the unfolding of an Indian modernity.

Saha's gradual transformation from a lower-caste *shudra* to a *bhadralok* scientist also entailed the building of the Institute named after him—the Saha Institute of Nuclear Physics (SINP). On the instrumentation side, he was successful, with the help of a network of people who respected him, like E.O. Lawrence and the Tatas and Birlas, in acquiring a cyclotron at SINP. In 1940, he introduced nuclear physics as a specialization for students getting a graduate degree in physics from Calcutta University. Saha also raised money from Nehru and saved enough funds to build a house in Calcutta, which was the ultimate sign of being a *bhadralok*.[127]

Thus, the seemingly rough transition from a lower *shudra* caste to a *bhadralok* scientist was complete. He played an active role in Indian politics after independence by becoming a member of the parliament, and had a different set of problems to grapple with.

Notes

1 Svein Rosseland. *Theoretical Astrophysics*. (Clarendon Press, 1936) xvi.
2 See also Endnote 83 in Chapter 2.
3 Kingsley Davis. *Human Society*. (New York: Macmillan Co. 1949) 96–117.
4 Scholars who have engaged with Saha in non-hagiographical ways are Abha Sur. *Dispersed Radiance*. (New Delhi: Navayana, 2011) and Robert Anderson. *Nucleus and Nation: Scientists, International Networks and Power in India*. (Chicago: University of Chicago Press, 2010); and about Saha's influence in the West see David H. DeVorkin. "Quantum Physics and the Stars (IV): Meghnad Saha's Fate." *Journal for the History of Astronomy* 25: 3 (August 1994) 155–188. https://doi.org/10.1177/0 02182869402500301.
5 Anderson. *Nucleus and Nation* 25.
6 Samarendra Nath Sen. *Professor Meghnad Saha: His Life, Work and Philosophy*. (Calcutta: Meghnad Saha 60th birthday committee, 1954).
7 Sur. *Dispersed Radiance* 70.
8 Anderson. *Nucleus and Nation* 26.
9 Ibid. 27.
10 Ibid.
11 Jagadish Chandra Bose. *Response in the Living and Non-living*. (Calcutta: Distant Mirror, 1902) first page.
12 Michael Silvestri. "The Sinn Fein of India: Irish Nationalism and the Policing of Revolutionary Terrorism in Bengal." *The Journal of British Studies* 39, 4 (2000) 454–486.
13 IB File no. 255/26, Kolkata (accessed July 2012).
14 *Ghadr* is Urdu for revolution.
15 Peter Hopkirk. *Like Hidden Fire: The Plot to Bring Down the British Empire*. (New York: Kodansha International, 1997) 82–84.
16 SINP archives, Folder 3404 (accessed, July 2012).
17 Enakshi Chatterjee and Santimay Chatterjee. *Meghnad Saha*. (National Book Trust, 2000) 98–106.
18 S.N. Bose stood first in this exam.
19 Meghnad Saha and S. Chakraborty. 1918. "On the Pressure of Light." *Journal of the Asiatic Society of Bengal* 14 (1918) 425.
20 Agnes Clerke. *Problems in Astrophysics*. (London: Adam & Charles Black, 1903).
21 Santimay Chatterjee (Ed.). *Collected Works of Meghnad Saha*, Vol 1. (Hyderabad: Orient Longman, 1982) vi–vii.
22 Ibid. 21–26.
23 The newspaper *The Statesman* was considered "the paper of record" of the high court. Published articles here were thought to be true, reliable, and authoritative by the intellectuals of India and also of Europe and North America. *The Statesman* had the status of *The Times* of London in the 1920s. The newspaper is still published in the present day and is the preferred news outlet for all intellectuals.
24 Ibid. 21–26.
25 Ibid.
26 Ibid.
27 Credit: Saha Institute of Nuclear Physics (SINP).

28 *The Calcutta University Commission Report*, Vol 8 (Calcutta University, 1919) 390–393.

29 Sorokin. *Social and Cultural Mobility* 138–140.

30 Abha Sur. *Dispersed Radiance.* 31, 69, 96–97, 104.

31 D.S. Kothari. "Meghnad Saha: 1893–1956." *Biographical Memoirs of the Fellows of the Royal Society* 5 (February 1960) 216–236. DOI: 10.1098/rsbm.1960.0017.

32 *Monthly notices.* 74 (1914) 425. *Journal of the R.A.S. of Canada*, 12 (1918) 357. Quoted in *Collected Scientific Works of Meghnad Saha.* (New Delhi: Council of Scientific and Industrial Research, Government of India, 1969).

33 Clerke. *Problems in Astrophysics* 51.

34 Lebedew. *Annalen Physik* 92 (1910) 411 Quoted in *Collected Scientific Works of Meghnad Saha.* (New Delhi: Council of Scientific and Industrial Research, Government of India, 1969).

35 Meghnad Saha. "On Radiation Pressure and the Quantum Theory, A Preliminary Note." *Astrophysical Journal* 50 (1919) 220 Quoted in *Collected Scientific Works of Meghnad Saha.*

36 Ibid. 21.

37 Ibid. 21.

38 Ibid.

39 Roger Stuewer. *The Compton Effect: Turning Point in Physics.* (Science History Publications, 1975). The Compton Effect was discovered and named after Arthur H. Compton in 1922.

40 Lebedew. (1910) 411. As quoted in *Collected Scientific Works.* Peter Lebedew worked in solar physics at the University of Moscow. E.F. Nichols and G.F. Hulls. "The Pressure due to Radiation. Second Paper." *Phys. Rev.* (Series 1) 17, 26 (July 1903) 26–50.

41 *Collected Scientific Works of Meghnad Saha* 21–22.

42 Ibid.

43 Ibid.

44 Dilip M. Salwi. *Meghnad Saha: Scientist with a Social Mission.* (New Delhi: Rupa, 2002).

45 Sumit Sarkar. *Modern India 1885–1947.* (Delhi: Macmillan, 1983).

46 Meghnad Saha. "Our National Crises." *Science and Culture* (1946) 253. As quoted in Santimay Chatterjee (eds.). *Collected Works of Meghnad Saha.* Vol.4 (1993) 308–315.

47 Abha Sur. *Dispersed Radiance* 77.

48 Salwi. *Meghnad Saha* 23.

49 Saha (1920) 472. Saha in his 1920 paper titled "Ionization in the Solar Chromosphere" in *Philosophical Magazine* argued that Lockyer's work was incomplete as it led to a contradictory hypothesis that the outer chromospheres was at a substantially higher temperature than the photosphere and the lower chromospheres; and that the temperature of the sun increases as one goes radially outwards, as quoted in Santimay Chatterjee, ed. *Collected Scientific Papers of Meghnad Saha.* (Calcutta: Saha Institute of Nuclear Physics, 1969).

50 Ibid.

51 Ibid.

52 D.S. Kothari. "Meghnad Saha: 1893–1956." *Biographical Memoirs of the Fellows of the Royal Society* 5 (1960) 216–236. Kothari was a Ph.D. student of Saha.

53 David DeVorkin. *Henry Norris Russell: Dean of American Astronomers.* (Princeton: Princeton University Press, 2000) 178.

54 *Collected Scientific Works of Meghnad Saha* (1969) 33–44; Kothari. "Meghnad Saha" 5.

55 Depending on direction of arrow.

56 Saha. (1920) 472. As quoted in *Collected Scientific Papers of Meghnad Saha.* Saha concluded this paper by remarking: "In conclusion, I beg to record my best thanks

to my students for their valuable help in the calculations, and to my friend Dr. J. C. Ghosh for revising the proofs."

57 G. Venkataraman. *Saha and his Formula*. (Hyderabad: Universities Press, 1995). For the value of U in equation (2) one can work backwards from the value of x, to find U.

58 I thank Professor David Cassidy for assisting me with the nuances of this calculation.

59 Saha. 40 (1920) 472 as quoted in *Collected Scientific Works of Meghnad Saha* 38–44.

60 SINP Archives, also at www.saha.ac.in/cs/archive.mns/picgallery/lc/lc3.html. The SINP claim from personal correspondence with the author in June 2012 that it is how-ever, not known whether Saha sent this application or not. Considering Saha was on a research leave in Europe from Calcutta and Saha had to grapple with "ivory tower" figures like Raman at Calcutta University and the lure of joining a new, well-funded Dacca University, it would not be surprising if Saha did send this letter.

61 *Science & Culture* archives.

62 Saha (1933) in *India and the World*. As quoted in Santimay Chatterjee. *Collected Works of Meghnad Saha. Vol. 2.* (Hyderabad: Orient Longman, 1986).

63 SINP archives. www.saha.ac.in/cs/archive.mns/picgallery/lc/lc6.html (accessed June 2012).

64 Aruparatana Bhattacharya. (trans. from Bengali). *Bijnani Meghanada Saha, jibana o sadhana* (trans. *The Scientist Meghnad Saha: Life and Works*). (Calcutta: Bestbooks, 1993) 84–85.

65 Mukherjee was also a patron of Indian science and Indian scientists. Financial sup-port for Indian science came from lawyers and philanthropists like Tarak Nath Palit, Nilratan Sircar, Rashbihari Ghosh, the Maharaja of Mysore, along with subsidies from the British government.

66 Saha and Gunther. 4 (1922) 97 As quoted in *Collected Scientific Papers of Meghnad Saha*.

67 *Collected Scientific Papers of Meghnad Saha* xi.

68 Nehru Archives: Saha Papers, New Delhi. Folder T5 (accessed July 2012).

69 Saha. (922) 605–622. As quoted in *Collected Works of Meghnad Saha* v.2, 3–20.

70 The word "post-doc" as an academic position was forged in the aftermath of World War I, especially in the United States due to efforts of Hale, Millikan, and Noyes. See David Kaiser. *Drawing Theories Apart: The Dispersion of Feynman Diagrams in Postwar Physics*. (Chicago and London: University of Chicago Press, 2005) 61–65.

71 1 lakh = 100,000.

72 Saha. (1932). "The Catastrophic Flood in Bengal and How They Can Be Combated." *Modern Review* 51, 163–168. As quoted in *Collected Works of Meghnad Saha* v.2, 21–26.

73 Within the structural framework of a nation there are many provinces inhabited by people who share a community sentiment with each other. Due to the close physical proximity, a feeling of belonging together develops, notwithstanding the fact that they are part of a larger community i.e., the nation.

74 Later, in 1947, before independence, when Muslim League demanded the entirety of Bengal to be handed over to the proposed Pakistan, there was a protest movement launched by the eminent *bhadraloks* of Bengal including Meghnad Saha, led by Dr. Shyamaprasad Mukherjee. As the western part of Bengal was inhabited mostly by Hindus, they argued that since India was being partitioned based on religion, and the areas with a Muslim majority were being given to Pakistan, then why should the western part of Bengal with a Hindu majority go to Pakistan? Saha and the others opposed the proposal of the Muslim League vehemently and ultimately the British government gave in and the western part of Bengal with Calcutta was saved from the proposed Pakistan. See for example, Shyamalesh Das. (1997) 70–80 (trans. from Bengali). Shyamalesh Das. [*Dr Shyamapasad: The Farsighted Statesman*] (in Bengali). (Calcutta: Shreebhumi Publishing Company, 1997).

75 Saha. (1922) 605–622. As quoted in *Collected Works of Meghnad Saha* v2, 7–21.

76 Saha. (1933) 237–259. As quoted in *Collected Works of Meghnad Saha*, v2, 11.

77 Ibid.

78 Ibid.

79 Ibid.

80 Saha. *Science & Culture* (1946). As quoted in *Collected Works of Meghnad Saha*, v2, 145–164.

81 S.B. Karmohapatro. *Meghnad Saha*. (Publications Division, Ministry of Information and Broadcasting, Government of India, 1997) 58–61. See further in earlier cited Bengali piece by Bhattacharya. 100–101 (trans. from Bengali).

82 *Science and Culture* (1935) 367. As quoted in *Collected Works* v.2, 182–189.

83 Ibid.

84 Ibid.

85 Ibid.

86 Ibid.

87 Anderson. *Nucleus and Nation* 64.

88 *Science and Culture* (1947). As quoted in *Collected Works* v.2, 43.

89 *Science and Culture* (1938) 546–547. As quoted in *Collected Works* v.4, 470–476.

90 Ibid.

91 Ibid.

92 Anthony Parel. Preface to the English translation. In Mohandas Karamchand Gandhi. *Hind Swaraj and Other Writings*. (Cambridge University Press, 1997) 164–172.

93 "Discussion with G. Ramachandran.". *Young India*, 13 November, 1924. As quoted in Judith Brown and Anthony Parel (eds.). *The Cambridge Companion to Gandhi*. (Cambridge University Press, 2011).

94 *Science and Culture* 1 (1936) 4.

95 *Science and Culture* 2 (1938) 4, 199.

96 Saha's 1939 foreword in the monograph "If the War Comes," in *Collected Works* Vol 3, 385. Agharkar. Saha gave a footnote here while speaking about Japan and said that while he appreciated the achievements of Japan, he unreservedly condemned Japan's aggressive policy towards China.

97 Agharkar (1939); Saha's foreword as quoted in *Collected Works* 3, 385–387.

98 Ibid.

99 Ibid.

100 Saha (1941). As quoted in *Collected Works* 3, 472–473.

101 Ibid.

102 Nehru papers correspondence, NMML Archives, New Delhi (accessed July 2012).

103 Jawaharlal Nehru. *Science and Culture* 8 (1942) 1.

104 *Science and Industry*. Address given by Nehru at the opening of the National Metallurgical Lab in Jamshedpur in 1950. Doc. 41, Nehru Memorial Museum and Library, New Delhi (accessed July 2012).

105 Speech at the inauguration of the Rare Earths Factory, Kerala, 1952. Ibid.

106 Baldev Singh, ed. *Nehru on Science and Society*. (New Delhi: Nehru Memorial Museum and Library, 1988) 10–13.

107 Benjamin Zachariah. *Developing India: An Intellectual and Social History, c. 1930–50. (*New Delhi: Oxford University Press, 2005).

108 Nehru 1960: 304 as quoted in Baldev Singh, *Nehru on Science and Society*.

109 The subcommittees were: Rural Marketing and Finance; River Training and Irrigation; Soil Conservation and Afforestation; Land Policy, Agricultural Labour and Insurance; Animal Husbandry and Dairying; Crop Planning and Production; Horticulture; Fisheries; Rural and Cottage Industries; Power and Fuel; Chemicals; Mining and Metallurgy; Engineering Industries including Transport Industries; Manufacturing Industries; Industries Connected with Scientific Instruments; Labour; Population; Trade; Industrial Finance; Public Finance; Currency and Banking;

Insurance; Transport Services; Communication Services; National Housing; Public Health; General Education; Technical Education and Sub-Committee on Woman's role in Planned Economy.

110   Meghnad Saha. "Science in Social and International Planning, With Special Reference to India." *Nature* 155 (February 24, 1945) 221–224 doi:10.1038/155221a0.

111   Karmohapatro. *Meghnad Saha* 91.

112   Meghnad Saha Papers, NMML, New Delhi (accessed July 2012). See also Santimay Chatterjee, and Jyotirmoy Gupta. *Meghnad Saha in Parliament*. (Calcutta: The Asiatic Society, 1993) 340–352.

113   Ibid.

114   Ibid.

115   Ibid.

116   Ibid.

117   Ibid.

118   His Russian memoirs were brought out in the form of a book by the publishing agency The Bookman in 1947. Meghnad Saha. *My Experiences in Soviet Russia*. (Calcutta: The Bookman, 1947).

119   *Collected Works* v.3, 407–469.

120   Ibid.

121   Ibid.

122   Ibid.

123   Vested interests such as becoming the first prime minister of independent India without any competition from iconic figures in science like Saha or in politics like Subhas Chandra Bose who had allegedly died in a "plane crash" in Taiwan in 1945.

124   Saha. *Nature* 155 (February 24, 1945) 221–224 doi:10.1038/155221a0.

125   Girjesh Govil. "Indian Science Congress and the Three Academies, 1914–35." In D.P. Chattopadhyaya (ed.). *History of Science, Philosophy and Culture in Indian Civilization* Vol XV, Part 4, *Science and Modern India: An Institutional History, c. 1784–1947*. (New Delhi: Centre for Studies in Civilizations, 2011) 146.

126   Reproduced from Somaditya Banerjee. "Meghnad Saha: Physicist and Nationalist." *Physics Today* 69, 8 (2016) 38 https://doi.org/10.1063/PT.3.3267 with the permission of the American Institute of Physics.

127   Robert Anderson/ *Nucleus and Nation* 133–148.

6   Reflections on *bhadralok* physics

In the spring of 1953, writes historian Danian Hu, a graduate student of history from California wrote to Albert Einstein and requested Einstein's thoughts on science in China. Responding to this request, Einstein said, "Development of Western Science is based on two great achievements: the invention of the formal logical system (in Euclidean geometry) by Greek philosophers, and the discovery of the possibility of finding out causal relationship by systematic experiment (Renaissance)."[1] Of these two prerequisites of modern science mentioned by Einstein, at least one was actually already available in ancient India—namely the robust system of logic in Vedic mathematics and computation. Thus, one might argue that this indigenous tradition made an impact on the process of subsequent successful reception and appropriation of European science by intellectuals of India.

This monograph conceptualized and analyzed a different and more recent indigenous system of knowledge production, called *bhadralok* physics, which was a specific brand of modern science that developed in late colonial India. Its emergence was made possible by the rise of a distinctively new social group of Indian intelligentsia. The *bhadraloks* were a relatively privileged stratum of intellectuals who deserved this name because they were, by Indian standards, well mannered, educated in a European style, respectable, and "characterized by a certain standard of personal and familial refinement."[2] The social esteem and self-identity of this group can be meaningfully compared to what in Wilhelmian Germany was called the *bildungsbürgertum*. As Gerhard Sonnert argues, the *bildungsbürgertum* were identified "by the certificates they had attained during the process of *Bildung*." The analogy with Germany illustrates that such a rise of a new intellectual class was not a uniquely Indian phenomenon, but part of a larger transnational trend—an epiphenomenon of modernization. Whereas the German intellectuals successfully employed the concept of *Kultur* to legitimize their collective claim for high social status in the German Empire, the cultural capital potentially available to Indian *bhadralok* intellectuals was more vulnerable, since they worked under an asymmetric power differential in the conditions of colonial domination. They used as their modus operandi other culturally specific tools described in this book as anticolonial nationalism, locally rooted *visvajaneen* cosmopolitanism and decolonization.

The specific group of *bhadralok* scientists examined in this monograph consisted mostly of early career scholars who had an educational background and postgraduate degrees in physics or mixed mathematics. It included several subsequently famous physicists, such as Satyendranath Bose, Chandrasekhara Venkata Raman, and Meghnad Saha, who were active from the early twentieth century onwards and lived to see India acquire the status of an independent country. They were educated in late-colonial India and received patronage from the developing institutions of Indian science, Calcutta University, Dacca University, the Indian Association for the Cultivation of Science (IACS) as well as from local patrons, such as the great educationist and legal luminary Sir Ashutosh Mukherjee. They were also able to visit and conduct research in Europe on postdoctoral or sabbatical fellowships and established contacts with scientists there, while receiving support and recommendations from some senior British and European colleagues.

Their unique accomplishments developed in the form of groundbreaking contributions to quantum physics—a revolutionary new field of international science at that time. The fact that India was able to develop advanced scientific research under colonial rule, along with the resulting major influence of a colony on the course of fundamental scientific development in the European metropole, constitutes an astonishing, and perhaps an unprecedented historical phenomenon that deserves special investigation. This monograph has undertaken such an investigation in the form of three related case studies.

Satyendranath Bose: Counting modernity

My first case study focused on the physicist Satyendranath Bose, who belonged to the *Kayastha* caste, which was a middle class in Indian perceptions compared to a lower-caste *shudra* or a higher caste of Brahmin. Bose's education was typical of a *bhadralok*, and he was also personally more reticent, withdrawn, and modest than the other two scientists studied here—Raman and Saha. One might argue that Bose was the most "Bhadra" among the three. This conclusion does not mean, however, that he was politically coy. Bose's anticolonial sentiment and sense of nationalism were, not untypically, awoken by Lord Curzon's partition of Bengal in 1905, causing a divide between the Muslim-dominated eastern and the Hindu-dominated western parts. Bose's example is particularly revealing as it shows that the nationalism of *bhadralok* scientists did not rule out internationalism. On the contrary, it was combined with a cosmopolitan outlook in a very specific way. I referred to Bose's characteristic fusion of nationalist aspirations and a cosmopolitan outlook with a deliberately self-contradictory term—"cosmopolitan nationalism," or, using Bengali words, *Visvajaneen Jatiyatabaad.* For Bose, orientation toward international science and collaboration with a world luminary such as Einstein also helped his nationalist agenda by providing him with an escape from intellectual dependence on the British Raj and its colonial hegemony.

Bose's collaboration with Einstein also revealed that the intellectual spectrum of a *bhadralok* physicist, even while pursuing international science, differed from that of a scientist from a European metropole. The analysis of his breakthrough contribution to quantum physics demonstrated that the breakthrough was cultur-ally and locally rooted in India. His peculiar adulation for Einstein as his intellec-tual *guru* reflected the Indian cultural nuance of the respect students held for their teachers, as enshrined in the ancient Indian philosophical texts. Bose's embrace of the concept of discontinuous light quanta was free from the typical British attachment to continuous Maxwellian electrodynamics and models of ether. It was much more unequivocal and wholehearted than the cautionary attitudes of the majority of European supporters of the idea, including even Einstein himself—the originator of the concept. Attachment to discontinuity that characterized Bose's thought in physics resonated with historical theories of his Indian mentor, chemist Prafulla Chandra Ray. The latter, in his *History of Hindu Chemistry*, conceptual-ized the entire span of Indian history as several periods of glorious past punctu-ated with discontinuous ruptures—the Aryan invasions from Central Asia, the Mughal Invasion from the North-West, and the subsequent British invasion from Europe. Bose's radically discontinuous worldview stretched beyond physics and the concept of the light quantum, reflecting also his political feelings regarding the fragmentation of Indian politics and the partition of Bengal.

I have highlighted that Bose's career and scientific success would not have been possible without a dialogue and mutual accommodation with his senior British colleagues and authorities, such as Jenkins and Hartog, who supported the devel-opment of Indian science. Several scholars, most notably Gyan Prakash,[3] Bernard Cohn,[4] Nicholas Dirks,[5] and most recently Shashi Tharoor,[6] have described the nature of British hegemony in South Asia. My analysis attenuates their conclu-sions by showing how it worked for the rise of modern science in India. It brings to light the complex processes of intercultural negotiations and collaborations, as well as subsequent cultural contingencies that were necessary for the produc-tion of indigenous scientific knowledge in the conditions of colonial society. Bose demonstrated in his physics that the relationship between "indigenous" and "European" could be complementary with regard to modern science, rather than antagonistic. His discovery of a novel kind of statistics en route to counting light quanta, and his concomitant introduction of the "indistinguishability" of quantum particles reflected the specific nature of Indian drive towards modernity.

C.V. Raman: Verifying modernity

Raman can also be characterized as a *bhadralok* scientist by education and occu-pation, even though he came from a rather privileged family. His brand of nation-alism was different from that of Bose or Saha, reflecting a strong influence of specific regional, in his case Southern Indian, identity. His gestures of showing fondness for the cultural traits of his country, such as wearing a turban, were not merely symbolic but also included genuine emotional attachment, as revealed by his feelings during the Nobel ceremony in 1930. Unlike Saha and Bose, who were both close to the Bengal Revolutionaries, Raman did not leave any evi-dence of association with nationalist organizations or individual revolutionaries. He seemed relatively comfortable operating within the modalities of the colonial society or, at the very least, was extremely successful at doing so. Although he lived and worked in Bengal for over two decades, he retained his strong identity as a southern Indian. His personal verbal quarrels with Saha, Jagadish Chandra Bose, and others in practice were a relatively common occurrence within that group, although in theory these quarrels were not quite compatible with the supposedly polite culture of the *bhadraloks*.

Raman's physics was fittingly more colonial in character and much less radical than Bose's. But it involved a fusion of indigenous and European traits, as it did for Bose. Raman contributed to quantum physics, but he was opposed to the most radical concept of the discontinuous light quantum. With regard to the structure of light, he was strongly attached to the classical wave theory, which at that time was also favored by most British physicists as well as many European ones, including Niels Bohr. Raman's fascination with the classical wave theory was built on his interest in the theory of music, especially in Indian musical instruments. Drawing on the theories and insights of the German polymath Hermann Helmholtz, Raman hoped to explain the acoustics of traditional Indian musical instruments—*ectara*, *sitar*, *tambura*, *veena*, and *tabla*. Raman's scientific project thus combined his attachment to European science with local intellectual traditions as a way to develop the specifically Indian brand of modernity.

Raman's main research accomplishment attests to the capacities of colonial science and its potential for competition with the metropole. His discovery provided an experimental verification of Smekal's theoretical insights and the subsequent Kramers–Heisenberg dispersion formula. The announcement of the Raman Effect made a major impact on European research agendas, both experimental and theoretical, and helped advance quantum mechanical theories of radiation and its interaction with matter. Raman's example refutes Orientalist stereotypes about a colonial scientist being incapable of doing original research, and indigenous knowledge always being a derivative of a metropolitan one. Science in colonial India also proved to be far subtler than what is typically implied by the center–periphery model of linear diffusion.[7]

Raman's case, as well as Bose's and Saha's, also helped to reveal limitations of the simple Indo-British hybridity model of knowledge production in the colonies. Their science included a much wider variety of multicultural influences than merely the cultures of the colonizer and the colonized. However, Raman's case depended crucially on the theories of musical instruments by Helmholtz more than on Rayleigh's ideas about light scattering, or later on, Bohr's stance regarding waves and quanta in radiation. Clearly, a confluence of multiple research traditions and intellectual ideas is noticeable in Indian science, which cannot justly be comprehended from the postcolonial theories of hybridity of Indo-British thought. One needs a multi-faceted theory or multidimensional matrix to account for such a multicultural synthesis—an idea that can be illustrated in the simplest case by a 2×2 matrix, for example:

$$\begin{pmatrix} \text{Indian}\,(i\text{l})\,\text{Russian}\,(r) \\ \text{British}\,(b)\,\text{German}\,(g) \end{pmatrix}$$

The cultural phenomenon of *bhadralok* physics requires a much more complex, multidimensional matrix, or rather a product of several such matrices reflecting its multi-hybridity. Individual elements of such a product of cultural matrix describe various combinations of cross-fertilization and can, correspondingly, describe the Bose–Einstein statistics, the Saha equation, and the Raman Effect.

Meghnad Saha: Applying modernity

Meghnad Saha was a contemporary of Raman. But unlike Raman, he came from the *shudra* caste—the lowest in the caste hierarchy of India. Despite strong discrimination and pressures from the upper sections of Indian society, Saha showed that it was possible to overcome the caste barrier through education and become recognized as a *bhadralok* scientist. His accomplishments as a physicist earned him respect and social mobility, with the benefit that he was no longer seen as a member of the lower caste, but rather as the scientist who developed an equation that was named after him. Saha's brand of nationalism was correspondingly more radical, as he closely associated himself with the Bengal Revolutionaries, particularly with organizations like *Anushilan Samity* and *Jugantar*, and with militant nationalists such as Jatindra Nath Mukherjee (also known as *Bagha Jatin*) and Pulin Das, whose methods included physical violence against the British colonial power. This risky involvement made Saha's academic career before India's independence much more difficult. However, international recognition for his research results helped ease his way through the tiers of academia.

Like Bose, Saha embraced the revolutionary quantum theory and its most radical concept of the light quantum, or the discontinuous structure of light, without reservation, which helped him make his own fundamental contribution to science. Saha's most important work in physics was the concept of selective radiation pressure and the analysis of its role in the relative distribution of elements in the solar atmosphere. He did this foundational work in astrophysics in 1919, when the discipline was still a nascent field. Before it was common for most European physicists (i.e., before the discovery of the Compton Effect in 1923), Saha employed the concept of energy-momentum of light as the basis for his astrophysical calculations of light pressure. He found that Maxwell's proof of the existence of light-pressure applied only to obstacles whose dimensions were larger than the wavelength of ordinary light. As atoms were much smaller than light waves, it followed that Maxwell's proof could not be applied to atoms.[8]

Thus, Saha had to turn to the new quantum theory in order to explain the tail of comets, and to remove the theoretical difficulty that classical approaches encountered. Saha explained the resultant pressure on the assembly of atoms using quantum theory of light, and gave a sounder basis regarding the nature of such pressure by formulating that pressure is exerted only if the atom can absorb the radiation. He concluded that "if white light is incident on Na-vapor, it is only the D1 and D2 light, which can exert pressure on the Sodium (Na) atoms. The action is therefore *Selective*."

Though Raman, Saha, and Bose lived and worked in Calcutta during the zenith of their careers, there were not too many instances of the three intellectuals working together. On one occasion, Saha and Bose collaborated early in their careers and published in *Philosophical Magazine* in 1918 on the equation of state.[9] On the contrary, Raman's research interests were quite separate, and did not overlap much with Saha and Bose's academic interests. Bengali was the vernacular language for both Saha and Bose, but Raman's mother tongue was Tamil. Interestingly enough, despite the absence of close personal interaction and collaboration, these scientists still show enough commonality in their political and research agendas to make it possible to conceptualize them as representative of "*bhadralok* physics." These scientists came to acquire the social status of a *bhadralok* by virtue of their education, manners, and intellectual contributions, despite their otherwise different caste and social class background. And like the moving patterns of a kaleidoscope, *bhadralok* physics revealed several different ways in which these scientists pursued their search for a specifically Indian modernity.

With respect to its scientific content, *bhadralok* physics cannot be classified as either exclusively theoretical or experimental in nature. It combined theory and experiment, both wave and particle aspects of light, and, likewise, both tradition and modernity in its various articulations. Bose, for example, was initially engaged in theoretical physics. But later, after returning to India from his European sojourn, he started training students in X-ray diffraction techniques he learned from Paul Langevin and the De Broglie brothers in Paris. Raman was figuring out the theory behind Indian musical instruments early in his career, but eventually switched to light-scattering experiments. Meanwhile, Saha aspired to collaborate with experimentalists in Berlin in 1921 who could have helped him test and verify his crucial theoretical insight into physics—the ionization theory and the equation named after him.

In *bhadralok* physics, we also encounter a lot of embedded mathematics, partly because science was not strictly compartmentalized into distinct disciplines during India's colonial period, as mathematics and physics are today. Bose and Saha had to study "mixed mathematics" during their college days, and this study made their quantitative skills very sharp. Others, like the Indian mathematician Ramanujan from Madras, did not have any formal scientific training. It is beyond the scope of this book to incorporate him into the *bhadralok* framework, but Ramanujan's contributions to mathematics can be compared to those analyzed here. Similarly, Ramanujan was recognized for his intellectual contributions, and not because he came from a relatively modest social class.[10]

Understanding "*bhadralok* physics" and developing this concept have several important repercussions for the history of physics, South Asian history, and science studies. It helps conceptualize how an original line of research in modern science was capable of developing and operating in a colony far away from a European metropole. Unlike their mentors—Jagadish Chandra Bose, Prafulla Chandra Ray, and Ganesh Prasad—Bose, Saha, and Raman belonged to the first generation of Indian scientists who were raised, educated, and professionally trained in India proper, rather than in European universities. Therefore, their *bhadralok* approach to physics brought into modern science some uniquely Indian concerns and sensitivities. Their styles of research were free from colonial connotations, and particularly independent of the conservative Cambridge Maxwellian framework which tended to favor the industrial hands-on approach, and possessed a strong bias toward the wave-theory of light. At the same time, during their research activities we often observed a dialogue—as opposed to hegemony—between the *bhadralok* physicists and their British supervisors and administrators. As a result, the novel knowledge produced by *bhadralok* scientists combined indigenous and European traditions in a unique manner, and can be seen as a characteristic feature of colonial modernity in various manifestations.

Subsequently, this book's findings also revealed correspondences between the *bhadralok* scientists' revolutionary anti-colonial aspirations and their attachments to relativity and quantum physics as the most radical examples of the twentieth-century revolution in sciences. These correspondences help to explain why Bose, Saha, and Raman specifically singled out these fields of research, which otherwise seemed quite unconnected with Indian realities and priorities at that time. Counterintuitively, it so happened that Indian science achieved its most important international successes in quantum physics, including the Saha ionization equation (1921), the famous Bose–Einstein statistics (1924), and the discovery of the Raman effect (1928), which earned the first and only Nobel Prize awarded to a scientist working in India. In a sense, Indian science can be said to have achieved its internationally famous intellectual accomplishments while it was still at its formative stages before it became firmly institutionalized in the country and before India acquired national independence.

Saha and Raman both contributed significantly to the process of institution-building for modern scientific research starting around 1930, using the models of IACS at Calcutta and Indian Institute of Science at Bangalore. Especially after decolonization and the end of British rule in India, science became massively institutionalized in India in the "big science" way, as it was in the West. Yet, despite the apparent success of the Nehruvian paradigm of institutionalization, the greatest intellectual products of Indian science still to this day belong to the *bhadralok* period. South Asian scholars who work on the history of science have been focusing mainly on the Indian nuclear program and its symbolic power. The history of science in India and the Indian nuclear program appear to be practically synonymous to many observers.[11] Though I believe that studying the nuclear program is important, my monograph and its findings reveal that eventual Indian successes in its nuclear aspirations were rooted in the earlier development of *bhadralok* physics, thereby adding a new dimension to the existing scholarship on the history of science in South Asia.

Finally, the concept of *bhadralok* physics helps to bridge the gap that exists between South Asian studies, in which the *bhadralok* phenomenon was initially discussed, and the history of science with its primary focus on the international dimensions of physics. Often times an analysis of the rise of nationalism in South Asian history has disregarded the crucial role of science and scientists in South Asia. As this book demonstrates, aspirations for national independence and decolonization depended not only on the rise of modern nationalism, but also, in very important ways, on the project of developing modern science, typically studied in the field of history of science as part of the *bhadralok* identity. We need a dialogue and synergy between these fields of historical research, and I hope that this book's research findings will help advance scholarship toward that direction.

Notes

1 Danian Hu. *China and Albert Einstein: The Reception of the Physicist and his Theory in China 1917–1979*. (Cambridge: Harvard University Press, 2005) 1–46.
2 Amit Kumar Gupta. *Crises and Creativities: Middle-Class Bhadralok in Bengal, c. 1939–52*. (Orient Blackswan, 2009) 7.
3 Gyan Prakash. *Another Reason: Science and the Imagination of Modern India*. (Princeton, Princeton University Press, 1999).
4 Bernard S. Cohn. *Colonialism and Its Forms of Knowledge: The British in India*. (Princeton: Princeton University Press, 1996).
5 Nicholas B. Dirks. *Castes of Mind: Colonialism and the Making of Modern India*. (Princeton: Princeton University Press, 2001).
6 Shashi Tharoor. *Inglorious Empire: What the British Did to India*. (London: C. Hurst & Co., 2017).
7 George Basalla. "The Spread of Western Science." *Science* 156 (1967) 611–622; Deepak Kumar. *Science and the Raj, 1857–1905*. (New Delhi: Oxford University Press, 1995); Paolo Palladino and Michael Worboys. "Science and Imperialism." *Isis* 84 (1993) 91–102.
8 Unless it is assumed the tail is formed of cosmic dust which are particles of about 10^{-5} cm in dimensions. But observations of cometary spectra show that the tails are composed of gases.
9 S.N Bose. "On the Influence of the Finite Volume of Molecules on the Equation of State." *Philosophical Magazine* 6 (1918) 36, 199.
10 Robert Kanigel. *The Man Who Knew Infinity: A Life of the Genius Ramanujan*. (New York: Washington Square Press, 2016).
11 MIT historian Abha Sur is a notable exception. She has written about women in science during this period, especially in the context of Anna Mani and her experience in Raman's laboratory. There can be feminist critiques of "*bhadralok* physics" being a solely male-dominated enterprise. But considering that the literacy of women in India in 1918 stood at less than 1%, lack of women in the sciences is not surprising. See Abha Sur. *Dispersed Radiance: Caste, Gender and Modern Science in India*. (New Delhi: Navayana, 2011). For example, Raman's wife was a *bhadramahila* (female equivalent of a *bhadralok*) and during their stay in Calcutta used to socialize with various Bengali *bhadramahila*, while Raman was collaborating at IACS with his South Indian experimental group.

Bibliography

Archives and special collections

Archives and Library, Satyen Bose National Center for Basic Science (SNBCS), Kolkata, India.

Archives and Library, Bose Institute, Kolkata, India.

Archives and Library, Indian Association for the Cultivation of Science (IACS), Kolkata.

Archives and Library, Calcutta Mathematical Society.

Archives and Library, Ballygunge Institute, Kolkata.

Archives and Library, National Library (NL), Kolkata.

Archives and Library, St. Xavier`s College (SXC), Kolkata.

Archives and Library, Pune University, Maharashtra, India.

Archives and Library, Inter University Centre for Astronomy and Astrophysics (IUCAA), Pune.

Archives and Library, National Centre for Radio Astrophysics (NCRA), Pune.

Archives and Library of the Indian Institute of Science, Bangalore, India Archives and Library of the Raman Research Institute (RRI), Bangalore, India.

Archives and Library of the Saha Institute of Nuclear Physics (SINP), Kolkata, India.

Archives and Library of Asiatic Society, Kolkata, India.

Archives and Library of Indian Chemical Society, Kolkata, India.

Archives and Library of the Nehru Memorial Museum and Library (NMML), New Delhi.

Archives and Library of Harish Chandra Research Institute (formerly Mehta Research Institute, Allahabad, India.

Archives and Online Collections, Deutsches Museum, München, Germany.

Einstein Papers Project, Pasadena, California.

Indian Science News Association, Kolkata.

India Office Records and Private Papers, The British Library Archives, London, UK. Jewish National and University Library Jerusalem Archive.

West Bengal State Archives, Kolkata.

Special Branch Calcutta Police, Government of West Bengal, Kolkata .

National Archives, New Delhi.

Taltala Public Library, Kolkata.

Personal collections

Enakshi Chatterjee and late Santimay Chatterjee, Kolkata, India.

Amar Roy, University of Delhi, New Delhi, India.

Asim Kumar Ganguly, University of Calcutta, Kolkata, India.
Partha Ghose, Satyen Bose Center for Basic Science, Kolkata, India.
Falguni Sarkar. *S.N. Bose Biography Project*. http://www.snbose.org/.
Jagdish Mehra Collections, Oral Recordings, University of Houston Libraries, Houston, Texas.
Rajinder Singh, Germany.
Daniel Kennefick, Arkansas.
Michel Janssen, Minnesota.

Journal archives

Science and Culture
Current Science
Asiatic Society of Bengal

Newspapers

The Statesman
Hindustan Times
The Hindu
New York Medical Times

Interviews

Partha Ghose, Calcutta
Enakshi Chatterjee, Calcutta
Tripurari Prasad Sinha, Calcutta
Amarendra Nath Roy, New Delhi
Rajaram Nityananda, Pune
P.K. Chakraborty, Calcutta
Robert Anderson, Vancouver

Secondary sources

Abraham, Itty. *The Making of the Indian Atomic Bomb: Science, Secrecy and the Postcolonial State*. London: Zed Books, 1998.
Adas, Michael. *Machines as the Measure of Men: Science, Technology and Ideologies of Western Dominance*. Ithaca: Cornell University Press, 1989.
Anderson, Anthony. *The Raman Effect*. New York: Marcel Dekker, 1971.
Anderson, Benedict. *Imagined Communities: Reflections on the Origin and Spread of Nationalism*. London and New York: Verso, 1983.
Anderson, Robert. *Building Scientific Institutions in India: Saha and Bhabha*. Occasional Paper Series, No. 11, Centre for Developing-Area Studies. Montreal: McGill University, 1975.
Anderson, Robert. *Nucleus and Nation: Scientists, International Networks and Power in India*. Chicago: University of Chicago Press, 2010.
Anderson, Warwick. "Special issue: Postcolonial technoscience." *Social Studies of Science* 32, nos. 5–6 (2002): 643–658.

Arabatzis, Theodore. "The electron's hesitant passage to modernity 1913–1925." In M. Epple and F. Müller (eds.), *Science as Cultural Practice*. Akademie-Verlag, forthcoming.

Arabatzis, Theodore. *Representing Electrons: A Biographical Approach to Theoretical Entities*. Chicago: The University of Chicago Press, 2006.

Arnold, David. *Science, Technology and Medicine in Colonial India*. The New Cambridge History of India, III, book 5, Cambridge: Cambridge University Press, 2000.

Ashcroft, Bill, Gareth Griffiths, and Helen Tiffin. *Post-Colonial Studies: The Key Concepts*. London and New York: Routledge, 2000.

Baber, Zaheer. *The Science of Empire: Scientific Knowledge, Civilization and Colonial Rule in India*. New Delhi: Oxford University Press, 1998.

Baber, Zaheer. "Colonizing nature: Scientific knowledge, colonial power and the incorporation of India into the modern world system." *British Journal of Sociology* 52, no. 1 (2001): 37–58.

Bagchee, Moni. *Bijñānasādhaka Satyena Bosa*. Calcutta: Annapurna Publishing House, 1974.

Banerjee, Somaditya. "Satyen Bose: The unsung hero of India." Paper Presented at the Joint Atlantic Seminar for the History of Physical Sciences JASHOPS, Notre Dame, Indiana, February 4–6, 2005.

Banerjee, Somaditya. "Review of Robert Anderson, nucleus and nation: Scientists, international networks, and power in India." *Annals of Science* 71 (2011): 107–110.

Banerjee, Somaditya. "C.V. Raman and colonial physics: Acoustics and the quantum." *Physics in Perspective* 16, no. 2 (2014): 146–178.

Banerjee, Somaditya. "Meghnad Saha: Physicist and nationalist." *Physics Today* 69, no. 8 (2016): 38–44.

Banerjee, Somaditya. "Transnational quantum: Quantum physics in India through the lens of Satyendranath Bose." *Physics in Perspective* 18, no. 2 (2016): 157–181.

Basalla, George. "The spread of western science." *Science* 156 (1967): 611–622.

Bellenoit, Hayden. *Missionary Education and Empire in Late Colonial India 1860–1920*. London: Pickering & Chatto, 2007.

Beller, Mara. *Quantum Dialogue: The Making of a Revolution*. Chicago: University of Chicago Press, 2001.

Bhabha, Homi. *The Location of Culture*. London and New York: Routledge, 1994.

Bhagavan, Manu. *Sovereign Spheres: Princes, Education and Empire in Colonial India*. Oxford: Oxford University Press, 2003.

Bhagavantam, S. "Professor Chandrasekhara Venkata Raman." *Biographical Memoirs of Fellows of the Royal Society London* 17 (1972): 565–579.

Bhattacharya, Aruparatana. *Bijnani Meghanada Saha, jibana o sadhana*. Kolkata: Bestbooks, 1993.

Bhattacharya, Tithi. *Sentinels of Culture: Class, Education, and the Colonial Intellectual in Bengal 1848–85*. New York: Oxford University Press, 2005.

Biagioli, Mario. *Galileo Courtier: The Practice of Science in the Culture of Absolutism*. Chicago: University of Chicago Press, 1994.

Biswas, Arun Kumar. *History, Science and Society in the Indian Context*. Calcutta: The Asiatic Society, 2001.

Biswas, Arun Kumar. *Collected Works of Mahendralal Sircar, Eugene Lafont and the Science Movement 1860–1910*. Kolkata: Asiatic Society, 2003.

Blanpied, William. "Satyendranath Bose: Co-founder of quantum statistics." *American Journal of Physics* 40 (1972): 1212–1220.

Blanpied, William. "Pioneer scientists in pre-independence India." *Physics Today* 39 (1986): 36–44.

Blanpied, William. "Pioneer scientists in pre-independence India." *Physics Today* 39 (1986): 36–44.

Bohr, N. "On the constitution of atoms and molecules." Part I. *Philosophical Magazine* 26 (1913): 1–25.

Bohr, N. "On the quantum theory of radiation and the structure of the atom." *Philosophical Magazine* 30 (1915): 394–415.

Bohr, N., H. A. Kramers, and J. C. Slater. "The quantum theory of radiation." *Philosophical Magazine* 47 (1924a): 785–802. Page references to reprint in (Van der Waerden, 1968, pp. 159–176).

Born, Max. "Fourth Russian physicists conference." *Naturwissenschaften* 16 (1928): 741.

Born, Max and Pascual Jordan. "Zur quantenmechanik." *Zeitschrift für Physik* 34 (1925): 858–888.

Born, Max and Emil Wolf. *Principles of Optics: Electromagnetic Theory of Propagation, Interference and Diffraction of Light*. London: Pergamon Press, 1959.

Bose, Debendra Mohan, B. V. Subbarayappa, and S. N. Sen. *A Concise History of Science in India*. Delhi: Indian National Science Academy, 1970.

Bose, S. N. "On the deduction of Rydberg's law from the quantum theory of spectral emission." *Philosophical Magazine* 40 (1920): 619.

Bose, Satyendranath. "On the horpolhode." *Bulletin of the Calcutta Mathematical Society* 11 (1919): 21.

Bose, Satyendranath. "The stress-equations of equilibrium." *Bulletin of the Calcutta Mathematical Society* 10 (1919): 117.

Bose, Satyendranath. "Plancks Gesetz und Lichtquantenhypothese." *Zeitschrift für Physik* 26 (1924a): 178–181.

Bose, Satyendranath. "Wärmegleichgewicht und Strahungsfeld bei Anwesenheit von Materie." *Zeitschrift für Physik* 27 (1924b): 383–393.

Bose, Sugata and Kris Manjapra. *Cosmopolitan Thought Zones: South Asia and the Global Circulation of Ideas*. New York: Palgrave Macmillan, 2010.

Bourdieu, Pierre. *Distinction: A Social Critique of the Judgement of Taste*. London and New York: Routledge, 1984.

Brand, J. C. D. "The discovery of the Raman effect." *Recordings of the Royal Society* 43 (1989): 1–23.

Breckenridge, Carol, et al. *Orientalism and the Postcolonial Predicament: Perspectives on South Asia*. Philadelphia: University of Pennsylvania Press, 1993.

Brown, Melvyn. *Satyendranath Bose: A Life*. Calcutta: Annapurna Publishing House, 1974.

Buchwald, J. D. "Historical unity." *Isis* 78 (1987): 244–249.

Camilleri, K. "Constructing the myth if the Copenhagen interpretation." *Perspectives on Science* 17 (2009): 26–57.

Chakrabarti, Pratik. *Western Science in Modern India: Metropolitan Methods, Colonial Practices*. New Delhi: Permanent Black, 2004.

Chakrabarti, Pratik. "Review of John Lourdusamy, science and national consciousness in Bengal." *Medical History* 50, no. 3 (2006): 403–404.

Chakrabarty, Bidyut. *The Partition of Bengal and Assam, 1932–1947: Contour of Freedom*. London and New York: Routledge, 2004.

Chakrabarty, Dipesh. *Provincialising Europe: Postcolonial Thought and Historical Difference*. New Delhi: Oxford University Press, 2000.

Chatterjee, Enakshi and Santimay Chatterjee. *Bharatiya Bijnan- Uttoroner Kal*. Kolkata: Cambridge, India, 2003.

Chatterjee, Partha. *The Nation and Its Fragments: Colonial and Postcolonial Histories. (Princeton Studies in Culture/Power/History)*. Princeton: Princeton University Press, 1993.

Chatterjee, Partha. *Nationalist Thought and the Colonial World: A Derivative Discourse*. Minneapolis: University of Minnesota Press, 1993.

Chatterjee, Partha. *Our Modernity*. Senegal: CODESRIA-SEPHIS, 1997.

Chatterjee, Santimay, ed. *Collected Scientific Papers of Meghnad Saha*. Kolkata: Saha Institute of Nuclear Physics, 1969.

Chatterjee, Santimay and Enakshi Chatterjee. *Meghnad Saha: Scientist with a Vision*. New Delhi: National Book Trust, 1984.

Chatterjee, Santimay. *Collected Works of Meghnad Saha. Vols. 1, 2, 3, 4*. Hyderabad: Orient Longman, 1987.

Chatterjee, Santimay and Jyotirmoy Gupta. *Meghnad Saha in Parliament*. Calcutta: The Asiatic Society, 1993.

Chatterjee, Santimay, ed. *S.N. Bose: The Man and His Work: Collected Scientific Papers (Part 1)*. Calcutta: SNBCS, 1994.

Chatterjee, Santimay, ed. *S.N. Bose: The Man and His Work: Life, Lectures and Addresses, Miscellaneous Pieces (Part 2)*. Calcutta: SNBCS, 1994.

Chatterjee, Santimay. "Meghnad Saha and C. V. Raman: Fact and fiction." *Indian Physical Society Diamond Jubilee Number* (1995): 43–47.

Chatterjee, Santimay and Enakshi Chatterjee. *Satyendra Nath Bose*. Calcutta: National Book Trust, 2005.

Chattopadhyay, Swati. *Representing Calcutta: Modernity, Nationalism, and the Colonial Uncanny*. London and New York: Routledge, 2006.

Chaturvedi, Vinayak. ed. *Mapping Subaltern Studies and the Postcolonial*. London and New York: Verso, 2000.

Chibber, Vivek. *Postcolonial Theory and the Specter of Capital*. New York: Verso, 2013.

Clerke, Agnes. *Problems in Astrophysics*. London: Adam & Charles Black, 1903.

Cohn, Bernard. *Colonialism and Its Forms of Knowledge: The British in India*. Princeton: Princeton University Press, 1996.

Compton, A. H. "A quantum theory of the scattering of X-rays by light elements." *Physical Review* 21 (1923): 483–502.

Cooper, Frederick. *Colonialism in Question: Theory, Knowledge, History*. Berkeley: University of California Press, 2005.

Crepeau, John. "Loschmidt, Stefan and Stigler's law of eponymy." *Physics in Perspective* 11, no. 4 (2009): 357–378.

Darrigol, Olivier. *From c-Numbers to q-Numbers: The Classical Analogy in the History of Quantum Theory*. Berkeley: University of California Press, 1992.

Darrigol, Olivier. "The electrodynamic origins of relativity theory." *Historical Studies in the Physical and Biological Sciences* 26 no. 2 (1996): 241–312.

Darwin, Charles Galton. "A quantum theory of optical dispersion." *Nature* 110 (1922): 841–842.

Darwin, Charles Galton. "The sixth congress of Russian physicists." *Nature* 122 (1928): 630.

Das, Shyamalesh. [*Dr Shyamapasad: The Farsighted Statesman*] (in Bengali). Kolkata: Shreebhumi Publishing Company, 1997.

Das, Sitanshu. *Subhash Chandra Bose- A Political Biography*. New Delhi: Rupa, 2001.

Dasgupta, Deepanwita. "Stars, peripheral scientists and equations." *Physics in Perspective* 17, no. 2 (2015): 83–106.

Dasgupta, Subrata. *Jagadish Chandra Bose and the Indian Response to Western Science.* New York: Oxford University Press, 2000.

Dasgupta, Subrata. *The Bengal Renaissance: Identity and Creativity from Rammohun Roy to Rabindra Nath Tagore.* New Delhi: Permanent Black, 2007.

Dasgupta, Uma, ed. *Science and Modern India: An Institutional History c. 1784–1947.* New Delhi: Centre for Studies in Civilizations, 2011.

Davis, Kingsley. *Human Society.* New York: Macmillan Co, 1949.

Debye, Peter. "Die Konstitution des Wasserstoff-molekuls." *Sitzungsberichte der mathematisch-physikalischen Klasse der Kniglichen Bayerischen Akademie der Wissenschaften zu Munchen* (1915): 1–26.

Debye, Peter. "Zerstreuungen von Röngtenstrahlen nach der Quantentheorie." *Zeitschrift für Physik* 24 (1923): 161–166.

Deshmukh, Chintamani. *Homi Jehangir Bhabha.* Delhi: National Book Trust of India, 2003.

DeVorkin, David. *Henry Norris Russell: Dean of American Astronomers.* Princeton: Princeton University Press, 2000.

DeVorkin, David. "Quantum physics and the stars (IV): Meghnad Saha's fate." *Journal for the History of Astronomy* 25, no. 3 (1994): 155–188. https://doi.org/10.1177/002182 869402500301

Dirac, P. A. M. *Principles of Quantum Mechanics,* 3rd ed. Oxford: Oxford University Press, 1947.

Dirks, Nicholas. *Castes of Mind: Colonialism and the Making of Modern India.* Princeton: Princeton University Press, 2001.

Dresden, Max. *H.A. Kramers: Between Tradition and Revolution.* New York: Springer, 1987.

Drude, Paul. *Lehrbuch der Optik.* Leipzig: S. Hirzel, 1900; English transl. *The Theory of Optics,* trans. C. R. Mann and R. A. Millikan. New York: Longmans, Green, 1902.

Duara, Prasenjit. *Rescuing History from the Nation.* Chicago: University of Chicago Press, 1995.

Duara, Prasenjit. *Decolonization: Perspectives from Now and Then.* London and New York: Routledge, 2004.

Duncan, Anthony and Michel Janssen. "On the verge of *Umdeutung* in Minnesota: Van Vleck and the correspondence principle (Part One)." *Archive for History of Exact Sciences* 61 (2007): 553–624.

Duncan, Anthony and Michel Janssen. "Pascual Jordan's resolution of the conundrum of the wave particle duality of light." *Studies in History and Philosophy of Modern Physics* 39 (2008): 634–666.

Dutt, Krishna and Andrew Robinson. *Rabindranath Tagore: The Myriad Minded Man.* New York: St. Martin's Press, 1996.

Dutta, Mahadev. *Satyen Bose.* Calcutta: Bongiya Bijnan Parishad, 1995.

Eaton, Richard. *The Rise of Islam and the Bengal Frontier, 1204–1760.* Berkeley: University of California Press, 1996.

Einstein, Albert. "On a heuristic viewpoint on the production and transformation of light." *Annalen der Physik* 17, no. 6 (1905): 132–148.

Einstein, Albert. "Uber die Entwickelung unserer Anschauungen "uber das Wesen und die Konstitution der Strahlung. Deutsche Physikalische Gesellschaft." *Verhandlungen* 11 (1909b): 482–500.

Einstein, Albert. "Theorie der Opaleszenz von homogenen Flüssigkeiten und Flüssigkeitsgemischen in der Nähe des kritischen Zustandes." *Annalen der Physik* 33 (1910): 1275–1298.

Einstein, Albert. "Strahlungs-Emission und -Absorption nach der Quantentheorie." *Deutsche Physikalische Gesellschaft. Verhandlungen* 18 (1916a): 318–323. Reprinted in facsimile as Doc. 34 in (Einstein, 1987–2006, Vol. 6).

Einstein, Albert. "Zur Quantentheorie der Strahlung." *Physikalische Gesellschaft Zürich. Mitteilungen* 18 (1916b): 47–62. Reprinted as (Einstein, 1917) and (in facsimile) as Doc. 38 in (Einstein, 1987–2006, Vol. 6).

Einstein, Albert. "Zur Quantentheorie der Strahlung." *Physikalische Zeitschrift* 18 (1917): 121–128. Reprint of (Einstein, 1916b). English translation in (Van der Waerden, 1968, pp. 63–77).

Einstein, Albert and Paul Ehrenfest. "Quantentheorie des Strahlungsgleichgewichts." *Zeitschrift für Physik* 19 (1923): 301–306.

Einstein, Albert. "Quantentheorie des einatomigen idealen Gases Quantum theory of monatomic ideal gases." *Sitzungsberichte der Preussichen Akademie der Wissenschaften Physikalisch-Mathematische Klasse* (1924): 261–267.

Einstein, Albert. *The Collected Papers of Albert Einstein* ed. J. Stachel et al. 9 Vols. Princeton: Princeton University Press, 2006.

Elias, Norbert. *The Civilising Process.* Oxford: Basil Blackwell, 1978.

Fabelinskii, I. L. *Optika i Spectroscopiya* 55 (1983): 591.

Fabelinskii, I. L. "The discovery of combination scattering of light in Russia and India." *Physics-Uspekhi* 46 (2003): 1105–1112.

Fang, L. China and Albert Einstein. "'The reception of the physicist and his theory in China, 1917–1979 [book review]." *The China Journal* 55 (2006): 211–212.

Forman, Paul. "Scientific internationalism and the Weimar Physicists: The ideology and its manipulation in Germany after WWI." *Isis* 64 (1973): 151–180.

Forman, Paul. "*Kausalität, Anschaulichkeit,* and *Individualität,* or How Cultural Values Prescribed the Character and Lessons Ascribed to Quantum Mechanics." In Nico Stehr and Volker Meja (eds.), *Society and Knowledge,* 333–347. New Jersey: Transaction Books, 1984.

Foucault, Michel. "Truth and power." In Colin Gordon (ed.), *Power/ Knowledge: Selected Interviews and Other Writings, 1972–1977,* 109–133. New York: Pantheon, 1976.

Friedman, Robert Marc. *The Politics of Excellence: Behind the Nobel Prize in Science.* New York: Henry Holt and Company, 2001.

Galison, Peter. "Kuhn and Quantum Controversy." *The British Journal for the Philosophy of Science* 32, no. 1 (1981): 71–85.

Galison, Peter. *Einstein's Clocks, Poincare's Maps: Empires of Time.* New York: W. W. Norton, 2003.

Galison, Peter. *Image and Logic: A Material Culture of Microphysics.* Chicago: University of Chicago Press, 1997.

Gandhi, Mohandas Karamchand. *Collected Works of Mahatma Gandhi.* Delhi: Publications Division, 1958.

Gandhi, Mohandas Karamchand. *Hind Swaraj and Other Writings,* ed. Anthony Parel. Cambridge: Cambridge University Press, 1997.

Gandhi, Leela. *Postcolonial Theory: A Critical Introduction.* New York: Columbia University Press, 1998.

Ganeri, J. *The Lost Age of Reason.* New York: Oxford University Press, 2011.

Geddes, Patrick. *An Indian Pioneer of Science: The Life and Work of Sir Jagadis C. Bose*. London: Longman, Green and Co., 1920.

Ghose, Partha. "Bose statistics: A historical perspective." In C. K. Majumdar, P. Ghose, E. Chatterjee, S. Chatterjee, and S. Bandyopadhyay (eds.), *S. N. Bose: The 'Man and His Work*, 35–70. Calcutta: S.N. Bose National Centre for Basic Sciences. Reprinted in K. C. Wali (ed.). 2009. *Satyendra Nath Bose—His Life and Times: Selected Works with Commentary*, 296–331. Singapore: World Scientific Publishing, 1994.

Ghose, Partha. "S.N. Bose: The man." In Dr. Anwar Hossain (ed.), *Albert Einstein and S. N. Bose*, 122–135. Bangladesh: Bangladesh Academy of Sciences and Goethe-Institut, 2005.

Ghosh, Durba. "Terrorism in Bengal: Political violence in the interwar years." In Durba Ghosh and Dane Kennedy (eds.), *Decentering Empire: Britain, India and the Transcolonial World*, 1–406. New Delhi: Orient Longman, 2006.

Ghosh, S. C. "Calcutta University and Science" *Indian Journal of History of Science* 29, no.1 (1994): 49–61.

Gilbert, Irene. "The Indian academic profession: The origins of a tradition of subordination." *Minerva* 10 (1972): 384–441.

Glick, Thomas, ed. *The Comparative Reception of Relativity*. Kluewer Academic Publishers, 1987.

Goswami, Manu. *Producing India: From Colonial Economy to National Space*. Chicago and London: University of Chicago Press, 2004.

Govil, Girjesh. "Indian science congress and the three academies, 1914–35." In D. P. Chattopadhyaya (ed.), *History of Science, Philosophy and Culture in Indian Civilization*. Vol XV, Part 4, *Science and Modern India: An Institutional History, c. 1784–1947*. New Delhi: Centre for Studies in Civilizations, 2011.

Gramsci, Antonio. *Selections from the Prison Notebooks*, ed. and trans. Quintin Hoare and Geoffrey Nowell Smith. London: Lawrence and Wishart, 1971.

Guha, Ranajit. *A Subaltern Studies Reader, 1986–1995* Minneapolis: University of Minnesota Press, 1997.

Gupta, Chitrarekha. *The Kayasthas: A Study in the Formation and Early History of a Caste*, Calcutta: K.P. Bagchi, 1996.

Halder, Gopal. "Revolutionary terrorism." *Studies in Bengal Rennaisance*. 3rd ed. Bengal: National Council of Education, 2002.

Halhed, Nathaniel Brassey. *A Code of Gentoo Laws or, Ordinations of the Pundits* (1778). Cambridge: University Press, 2013.

Hardiman, David and Projit Bihari Mukharji. *Medical Marginality in South Asia: Situating Subaltern Therapeutics*. London and New York: Routledge, 2012.

Harding, Sandra. *Is Science Multicultural? Postcolonialisms, Feminisms and Epistemologies*. Bloomington: Indiana University Press, 1998.

Heisenberg, Werner. "Über die quantentheoretische Umdeutung kinematischer und mechanischer Beziehungen." *Zeitschrift für Physik* 33 (1925): 879–893. Page references to English translation in (Van der Waerden, 1968, pp. 261–276).

Helmholtz, Hermann von. *On the Sensations of Tone as a Physiological Basis for the Theory of Music*. London: Longmans Green & Co, 1875.

Hendry, John. "Bohr-Kramers-Slater: A virtual theory of virtual oscillators and its role in the history of quantum mechanics." *Centaurus* 25 (1981): 189–221.

Hendry, John. *The Creation of Quantum Mechanics and the Bohr-Pauli Dialogue*. Dordrecht: Reidel, 1984.

Herzfeld, K. F. "Versuch einer quantenhaften Deutung der Dispersion." *Zeitschrift für Physik* 23 (1924): 341–360.

Hoffman, Banesh. *Albert Einstein: Creator and Rebel*. New York: Penguin Books, 1972.

Hopkirk, Peter. *Like Hidden Fire: The Plot to Bring Down the British Empire*. New York: Kodansha International, 1997.

Hu, D. *China and Albert Einstein*. Cambridge: Harvard University Press, 2005.

Hu, D. "The reception of relativity in China." *Isis* 98 (2007): 539–557.

Indian Science Congress Association. *The Shaping of Indian Science: Indian Science Congress Association Presidential Addresses, 1914–1947*. Hyderabad: Universities Press, 2003.

Jammer, Max. *The Conceptual Development of Quantum Mechanics*. New York: McGraw-Hill, 1966.

Jammer, Max. *The Philosophy of Quantum Mechanics: The Interpretation of Quantum Mechanics in Historical Perspective*. New York: Wiley, 1974.

Janssen, Michel. "Drawing the line between kinematics and dynamics in special relativity." *Studies in History and Philosophy of Modern Physics* 40 (2009): 26–52.

Jayaraman, A. and A. K. Ramdas. "Chandrasekhara Venkata Raman." *Physics Today* 41 (1988): 56.

Johnston, Marjorie, ed. *The Cosmos of Arthur Holly Compton*. New York: Alfred A. Knopf, 1967.

Jordan, Pascual. "Early years of quantum mechanics: Some reminiscences." In J. Mehra (ed.), *The Physicist's Conception of Nature*, 294–299. Dordrecht: Reidel, 1973.

Kaiser, David. *Drawing Theories Apart: The Dispersion of Feynman Diagrams in Postwar Physics*. Chicago and London: University of Chicago Press, 2005.

Karmohapatro, S. B. "Meghnad Saha." Publications Division, Ministry of Information & Broadcasting, Government of India, 1997.

Kaviraj, Sudipta. "An outline of a revisionist theory of modernity." *Archives European Sociology* 46, no. 3 (2005): 497–526.

Keswani, G. H. *Raman and His Effect*. New Delhi: National Book Trust of India, 1980.

Klein, Martin, ed. *The Collected Papers of Albert Einstein*. Vol. 3. Princeton: Princeton University Press, 1993.

Klein, M. J. "Max planck and the beginnings of quantum theory." *Archive for History of Exact Sciences* 1 (1962): 459–479.

Klein, M. J. "Planck, entropy, and quanta, 1901–1906." *The Natural Philosopher* 1 (1963): 83–108.

Klein, M. J. "Thermodynamics and quanta in Planck's work." *Physics Today* 19, no. 11 (1966): 23–32.

Klein, M. J. "Thermodynamics in Einstein's thought." *Science* 157 (1967): 509–516.

Klein, M. J. "The first phase of the Bohr-Einstein dialogue." *Historical Studies in the Physical Science* 2 (1970): 1–39.

Klein, M. J. "Essay review of the quantum discontinuity." *Isis* 70 (1979): 430–434.

Kojevnikov, Alexei. "Einstein's fluctuation formula and the wave-particle duality." In Yuri Balashov and Vladimir Vizgin (eds.), *Einstein Studies in Russia. Einstein Studies*. Vol. 10, 181–228. Boston: Basel, Birkhäuser, 2002.

Kopf, David. *British Orientalism and the Bengal Renaissance: The Dynamics of Indian Modernization 1773–1835*. Berkeley and Los Angeles: University of California Press, 1969.

Kothari, D. S. "Meghnad Saha: 1893–1956." *Biographical Memoirs of the Fellows of the Royal Society* 5 (1960): 216–236.

Kragh, Helge. *Quantum Generations: A History of Physics in the Twentieth Century.* Princeton: Princeton University Press, 1999.

Kramers, H. A. and W. Heisenberg. "Über die Streuung von Strahlung durch Atome."*Zeitschrift für Physik* 31 (1925): 681–707.

Krishnan, R. S. *Raman Effect: Discovery and After.* New Delhi: National Physical Laboratory, 1978.

Krishnan, R. S. and R. K. Shankar. "Raman effect: History of the discovery." *Journal of Raman Spectroscopy* 10, no. 198 (1981): 1–8.

Kuhn, T. S. *The Essential Tension.* Chicago: The University of Chicago Press, 1977.

Kuhn, T. S. *The Structure of Scientific Revolutions* (1962). Chicago: The University of Chicago Press, 2012.

Kumar, Deepak. *Science and the Raj, 1857–1905.* New Delhi: Oxford University Press, 1995.

Kumar, Deepak. "*The Making of the Indian Atomic Bomb: Science, Secrecy, and the Postcolonial State.* Itty Abraham." *Isis* 92, no. 1 (2001): 213–214.

Kumar, Prakash. *Indigo Plantations and Science in Colonial India.* Cambridge and New York: Cambridge University Press, 2012.

Ladenburg, Rudolf. "Uber die Dispersion des leuchtenden Wasserstoffs." *Physikalische Zeitschrift* 9 (1908): 875–878.

Ladenburg, Rudolf and Stanislaw Loria. "Uber die Dispersion des leuchtenden Wasserstoffs." *Deutsche Physikalische Gesellschaft. Verhandlungen* 10: 858–866. Reprinted in *Physikalische Zeitschrift* 9 (1908): 875–878.

Ladenburg, Rudolf. "Die quantentheoretische Deutung der Zahl der Dispersionselektronen." *Zeitschrift für Physik* 4 (1921): 451–468. Page references are to English translation in (Van der Waerden, 1968, pp. 139–157).

Ladenburg, Rudolf and Fritz Reiche. "Absorption, Zerstreuung und Dispersion in der Bohrschen Atomtheorie." *Die Naturwissenschaften* 11 (1923): 584–598.

Ladenburg, Rudolf and Fritz Reiche. "Dispersionsgesetz und Bohrsche Atomtheorie." *Die Naturwissenschaften* 12 (1924): 672–673.

Ladenburg, Rudolf. "Die quantentheoretische Dispersionsformel und ihre experimentelle Prüfung." *Die Naturwissenschaften* 14 (1926): 1208–1213.

Ladenburg, Rudolf. "Untersuchungen über die anomale Dispersion angeregter Gase. I. Teil. Zur Prüfung der Quantentheoretischen Dispersionsformel." *Zeitschrift für Physik* 48 (1928): 15–25.

Landsberg, G. and L. Mandelstam. "Eine neue Erscheinung bei der Lichtzerstreuung in Krystallen." *Die Naturwissenschaften* 16 (1928): 557–558.

Long, Derek. *The Raman Effect: A Unified Treatment of the Theory of Raman Scattering by Molecules.* England: John Wiley 2002.

Loomba, Ania. *Colonialism/Postcolonialism.* London and New York: Routledge, 1998.

Lorentz, H. A., A. Einstein, H. Minkowski, and H. Weyl. *The Principle of Relativity, A Collection of Original Memoirs on the Special and General Theory of Relativity* (with notes by A. Sommerfeld), trans. W. Perrett and G. B. Jeffery. London: Methuen & Co., 1923.

Loudon, Rodney. *The Quantum Theory of Light.* 3rd ed. New York: Oxford University Press, 2000.

Lourdusamy, John. *Science and National Consciousness in Bengal, 1870–1930.* New Delhi: Orient Longman, 2004.

Ludden, David. ed. *Reading Subaltern Studies: Critical History, Contested Meaning, and the Globalisation of South Asia.* New Delhi: Permanent Black, 2001.

MacKinnon, E. M. "Heisenberg, models, and the rise of matrix mechanics." *Historical Studies in the Physical Sciences* 8 (1977): 137–188.

MacKinnon, E. M. *Scientific Explanation and Atomic Physics*. Chicago: University of Chicago Press, 1982.

Macleod, Roy. "On visiting the 'moving metropolis': Reflection on the architecture of imperial science." *Historical Records of Australian Science* 5, no. 3 (1982): 217–249.

Majumdar, Ramesh Chandra. *The Sepoy Mutiny and the Revolt of 1857*. Calcutta: Firma K.L. Mukhopadhyay, 1963.

Majumdar, Romesh Chandra. "The Genesis of Extremism." In Chittabrata Palit (ed.), *Studies in Bengal Renaissance*. 3rd ed., 187–202. Bengal: National Council of Education, 2002.

Majumdar, Sisir. "Rabindranath's thoughts on science." *Frontier* 44 (2011): 1–5.

Mallick, D. C. V. and S. Chatterjee. *Kariamakkam Srinivasa Krishnan: His Life and Work*. Hyderbad: Universities Press, 2011.

Markovits, Claude. "How British was British India? Recovering the cosmopolitan dimension in the British-Indian colonial encounter." *Jahrbuch fur Europaische Uberseegeschichte* 10 (2010): 67–91.

McCormach, Russell. "Lorentz and the electromagnetic view of nature." *Isis* 61 (1970): 459–497.

McGuire, John. *Making of a Colonial Mind: A Quantitative Study of the Bhadralok in Calcutta, 1857–1885*. Canberra: Australian National University Press, 1983.

Mehra, Jagdish. "Satyendra Nath Bose." *Biographical Memoirs of Fellows of the Royal Society* 21 (1975): 117–154.

Mehra, Jagadish and Helmut Rechenberg. *The Historical Development of Quantum Theory*. Vols. 1–6. New York: Springer, 1982–2001.

Metcalf, Barbara and Thomas R. Metcalf. *A Concise History of Modern India*. Cambridge: Cambridge University Press, 2006.

Mill, James. *History of British India*. Vol. 1. London: Baldwin, Cradock and Joy, 1817.

Mill, James. *The History of British India*. London: J. Madden, 1848.

Misra, B. B. *The Indian Middle Classes*. London: Oxford University Press, 1961.

Misra, B. B. *The Indian Middle Classes*. London: Oxford University Press, 1965.

Mitra, Peary Chand and Gauranga Gopal Sengupta. *A Biographical Sketch of David Hare*. Calcutta: Best Books, 1979.

Nandy, Ashis. "Defiance and conformity in science: The identity of Jagadish Chandra Bose." *Science Studies* 2, no. 1 (1972): 31–85.

Nandy, Ashis. "From outside the imperium: Gandhi's cultural critique of the West." In Ashis Nandy (ed.), *Traditions, Tyranny, and Utopias: Essays in the Politics of Awareness*, 127–162. Delhi: Oxford University Press, 1987.

Nandy, Ashis. *Science, Hegemony and Violence: A Requiem for Modernity*. Delhi: Oxford University Press, 1990.

Nandy, Ashis. *Alternative Sciences: Creativity and Authenticity in Two Indian Scientists*. New Delhi: Oxford University Press, 1995.

Narlikar, Jayant. *An Introduction to Cosmology*. 3rd ed. London: Cambridge University Press, 2002.

Nehru, Jawaharlal. *The Unity of India: Collected Writings 1937-40*, London: Lindsay Drummond, 1941.

Nehru, Jawaharlal. *The Discovery of India*. New Delhi: Oxford University Press, 1946.

Osborne, Michael A., ed. "Focus Section: The Social History of Science, Technoscience and Imperialism." *Science, Technology and Society* 4, no. 2 (1999): 161–170.

Pais, Abraham. "Einstein on particles, fields, and the quantum theory." In M. Wolf (ed.), *Some Strangeness in Proportion*, 197–251. Reading: Addison Wesley, 1980.

Pais, Abraham. *Subtle Is the Lord*: *The Science and Life of Albert Einstein*. Oxford: Oxford University Press, 1982.

Palit, Chittabrata. *Science and Nationalism in Bengal: 1876–1947*. Kolkata: Institute of Historical Studies, 2004.

Palladino, Paolo and Michael Worboys. "Science and imperialism." *Isis* 84 (1993): 91–102.

Pandey, Gyanendra. *Routine Violence: Nations, Fragments, Histories*. Stanford: Stanford University Press, 2006.

Parameswaran, Uma. *C.V. Raman: A Biography*. New Delhi: Penguin Books, 2011.

Pauli, Wolfgang. "Über das thermische Gleichgewicht zwischen Strahlung und freien Elektronen." *Zeitschrift fur Physik* 18 (1923): 272–286.

Peabody, N. "After colonialism: Imperial histories and postcolonial displacements: Orientalism and the postcolonial predicament: Perspectives on South Asia [book review]." *American Ethnologist* 24 (1997): 212–213.

Perkovich, George. *India's Nuclear Bomb: The Impact on Global Proliferation*. Berkeley: University of California Press, 1999.

Phalkey, Jahnavi. "Not only smashing atoms: Nuclear physics at the University Science College, Calcutta, 1938–1948." In Uma Dasgupta (ed.), *Science and Modern India: An Institutional History c. 1784–1947*, 1057–1094. New Delhi: Pearson Education, 2010.

Philip, Kavita. *Civilizing Natures: Race, Resources and Modernity in Colonial South India*. New Brunswick and New Jersey: Rutgers University Press, 2004.

Pickering, Andrew. *Constructing Quarks: A Sociological History of Particle Physics*. Chicago: The University of Chicago Press, 1984.

Pollock, Sheldon. *The Language of the Gods in the World of Men: Sanskrit, Culture and Power in Premodern India*. Berkeley: University of California Press, 2009.

Pradhan, Ganesh. *Lokamanya Tilak*. India: National Book Trust, 1994.

Prakash, Gyan. *Another Reason, Science and the Imagination of Modern India*. Princeton: Princeton University Press, 1999.

Prasad, Baini. *The Progress of Science in India During The Past Twenty-Five Years*. Calcutta: The Indian Science Congress Association, 1938.

Prasad, B. N. "Obituary: Prof Ganesh Prasad, his life and work." *Science and Culture* I (1935): 142–145.

Prasad, Parmanand. *The Damodar Valley Corporation: A Brief Study*. New Delhi: Indian Institute of Public Administration, 1963.

Pratt, Mary Louise. *Imperial Eyes: Travel Writing and Transculturation*. London and New York: Routledge, 1991.

Price, Katy. *Loving Faster than Light*: *Romance and Readers in Einstein's Universe*. Chicago: University of Chicago Press, 2012.

Pringsheim, Peter. "Der Ramaneffekt, ein neuer von C.V. Raman entdeckter Strahlungseffekt." *Nature* 31 (1928): 597–601.

Pyenson, Lewis. "Cultural imperialism and exact sciences revisited." *Isis* 84 (1993): 103–108.

Raina, A. K. and B. N. Patnaik. *Science and Tradition*. Shimla: Indian Institute of Advanced Study, 2000.

Raina, Dhruv and S. Irfan Habib. *Domesticating Modern Science: A Social History of Science and Culture in Colonial India*. New Delhi: Tulika Books, 2004.

Raj, Kapil. *Relocating Modern Science: Circulation and the Construction of Knowledge in South Asia and Europe, 1650–1900*. London: Palgrave Macmillan, 2007.

Ramachandran, G. N. "Professor Raman: The Artist-Scientist." *Current Science* 40 (1971): 212–214. (the reference below is separate)

Raman, Chandrasekhara Venkata. "Unsymmetrical diffraction bands due to a rectangular aperture." *Philosophical Magazine* 6, no. 12 (1906): 494–498.

Raman, Chandrasekhara Venkata. "The Ectara." *The Journal of the Indian Mathematical Club* (1909): 170–175.

Raman, Chandrasekhara Venkata. "The colour of the Sea." *Nature* 108 (1921): 367.

Raman, Chandrasekhara Venkata and B. B. Ray. "On the transmission colours of sulphur suspensions." *Proceedings of the Royal Society of London.* A 100 (1921): 102–109.

Raman, Chandrasekhara Venkata. "Transparency of liquids and colour of the sea." *Nature* 110 (1922): 280.

Raman, Chandrasekhara Venkata. "A new radiation." *Indian Journal of Physics* 2 (1928): 387–398.

Raman, Chandrasekhara Venkata. "A classical derivation of the Raman effect." *Indian Journal of Physics* 3 (1929): 357–369.

Raman, Chandrasekhara Venkata. "Nobel address: Molecular scattering of light." *Indian Journal of Physics* 6 (1931): 263–273.

Raman, Chandrasekhara Venkata and S. Bhagavantam. "Experimental proof of the Spin of the Photon." *Indian Journal of Physics* 6 (1931): 355.

Raman, Chandrasekhara Venkata and S. Bhagavantam. "Experimental proof of the spin of the photon." *Nature* 129 (1932): 22–23.

Raman, Chandrasekhara Venkata. *Books That Have Influenced Me: A Symposium.* Madras: G. A. Natesan & Co., 1947.

Raman, Chandrasekhara Venkata. *New Physics: Talks on Aspects of Science.* Freeport: Books for Libraries Press, 1951.

Ramdas, L. A. "Raman effect in gases and vapours." *Indian Journal of Physics* 3 (1928): 131.

Ramsheshan, S. and C. Ramachandran, eds. *C.V. Raman: A Pictorial Biography.* Bangalore: The Indian Academy of Sciences, 1988.

Ray, Kamalesh. *The Life and Work of Meghnad Saha.* New Delhi: National Council of Educational Research and Training, 1968.

Ray, Prafulla Chandra. *A History of Hindu Chemistry.* Vols. 1–2. Calcutta: Chuckervertty, Chatterjee & Co, 1902–08.

Ray, Prafulla Chandra. *Life and Experiences of a Bengali Chemist.* Calcutta: Chuckervertty, Chatterjee & Co., Ltd., 1935.

Raychaudhury, Tapan. *Europe Reconsidered: Perceptions of the West in 19th Century Bengal.* Delhi, New York: Oxford University Press, 2002.

Rayleigh, John William Strutt. "Colours of sea and sky." *Royal Institution Proceedings* (February 25, 1910); *Nature*, LXXXIII (1910): 48.

Reiche, F. and W. Thomas. "Uber die Zahl der Dispersionselektronen, die einem station"aren Zustand zugeordnet sind." *Zeitschrift fur Physik* 34 (1925): 510–525.

Ringer, Fritz. *The Decline of the German Mandarins: The German Academic Community, 1890-1933.* Hanover: Wesleyan, 1990.

Rocard, Yves. "Les nouvelles radiations diffusées." *Comptes Rendus* 186 1928: 1107–1108.

Rosseland, Svein. *Theoretical Astrophysics.* Oxford: The Clarendon Press, 1936.

Saha, Meghnad. "On Maxwell's stresses." *Philosophical Magazine*, Sr. VI, 33 (1917): 256.

Saha, Meghnad and S. Chakraborty. "On the pressure of light." *Journal of the Asiatic Society of Bengal* 14 (1918): 425.

Saha, Meghnad. "On radiation pressure and the quantum theory, a preliminary note." *Astrophysical Journal* 50 (1919): 220.

Saha, Meghnad. "Ionization in the solar chromosphere." *Philosophical Magazine* VI, 40 (1920): 472.

Saha, Meghnad and Satyendranath Bose. "On the equation of state." *Philosophical Magazine* 6, no. 39 (1920): 456.

Saha, Meghnad and Satyendranath Bose. *The Principle of Relativity*. Calcutta: Calcutta University Press, 1920.

Saha, Meghnad. "Versuch einer Theorie der physikalischen Erscheinungen bei hohen Temperaturen mit Anwendungen auf die Astrophysik." *Zeitschrift für Physik* 6 (1921): 40.

Saha, Meghnad. "The great flood in northern Bengal." *Modern Review* 32 (1922): 605–622.

Saha, Meghnad and P. Gunther. "On the ionization of gases by heat." *Journal, Department of Science, Calcutta University* 4 (1922): 97.

Saha, Meghnad with B. N. Srivastava. *A Textbook of Heat Including Kinetic Theory of Matter, Thermodynamics, Statistical Mechanics and Theories of Thermal Ionization*. Allahabad: Indian Press, 1931.

Saha, Meghnad. "The catastrophic flood in Bengal and how they can be combated." *Modern Review* 51 (1932): 163–168.

Saha, Meghnad. "Need for a hydraulic research laboratory." Sir P.C. Ray's 70th Birthday Commemoration volume, *Indian Chemical Society Calcutta* (1933): 237–259.

Saha, Meghnad. "Obituary: Dr. Brühl." *Science and Culture* I (1935): 2.

Saha, Meghnad. "Progress of physics in India during the past twenty-five years." In B. Prashad (ed.), *The Progress of Science in India during the Past Twenty Years*, 674. Calcutta: Indian Scient Congress Association, 1938.

Saha, Meghnad. *My Experiences in Soviet Russia*. Calcutta: The Bookman, 1947.

Said, Edward. *Orientalism*. New York: Vintage, 1979.

Sarkar, Sumit. *The Swadeshi Movement in Bengal, 1903–1908*. Delhi: People's Publishing House, 1973.

Sarkar, Sumit. *Modern India 1885-1947*. Delhi: Macmillan, 1983.

Schiebinger, Londa. "Forum Introduction: The European Colonial Science Complex." *Isis* 96, no. 1 (2005): 52–55.

Segre, Emilio. *From X-rays to Quarks: Modern Physicists and Their Discoveries*. Mineola: Dover Publications, 2007.

Sen, Samarendra Nath. *Professor Meghnad Saha: His Life, Work and Philosophy*. Calcutta: Meghnad Saha 60th Birthday Committee, 1954.

Sengupta, Nitish. *The History of the Bengali Speaking People*. New Delhi: UBS Publishers 2002.

Seth, Suman. *Crafting the Quantum: Arnold Sommerfeld and the Practice of Theory, 1890–1926*. Cambridge, MA: MIT Press, 2010.

Shapin, Steven and Simon Schaffer. *Leviathan and the Air-Pump*. Princeton: Princeton University Press, 1989.

Shenstone, A. G. "Ladenburg, Rudolf Walther." In Charles Gillispie (ed.), *Dictionary of Scientific Biography*. Vol. VII, 552–556. New York: Charles Scribner's Sons, 1973.

Shah, K. T. *Handbook of National Planning Committee*. Bombay: Vora & Co., 1946.

Shils, Edward. "The academic profession in India." *Minerva* 7, no. 3 (1969): 387.

Silvestri, Michael. "The Sinn Fein of India": Irish nationalism and the policing of revolutionary terrorism in Bengal." *The Journal of British Studies* 39 (2000): 454–486.

Singh, Baldev. *Nehru on Science and Society*. New Delhi: Nehru Memorial Museum and Library 1988.

Singh, Jagjit. *Some Eminent Indian Scientists*. Ministry of Information and Broadcasting: Government of India Publications, 1966.

Singh, Rajinder. "Arnold Sommerfeld, The supporter of Indian physics in Germany." *Current Science* 81 (2001): 1489–1494.

Singh, Rajinder. "C.V. Raman and the discovery of the Raman effect." *Physics in Perspective* 4 (2002): 399–420.

Singh, Rajinder. "The story of C.V. Raman's resignation from the Fellowship of the Royal Society of London." *Current Science* 83 (2002): 1157–1158.

Singh, Rajinder. *Nobel Laureate C. V. Ramans's Work on Light Scattering: Historical Contributions to a Scientific Biography*. Hamburg: Logos Verlag, 2004.

Sinha, P. *Nineteenth Century Bengal: Aspects of Social History*. Calcutta: Firma K.L. Mukhopadhyaya, 1965.

Sinha, Jagadish. "Technology for national reconstruction: The national planning committee, 1938-49." In Roy MacLeod and Deepak Kumar (eds.), *Technology and the Raj*, 250–264. New Delhi, Thousand Oaks, London: Sage Publications, 1995.

Smekal, Adolf. "Zur Quantentheorie der Dispersion." *Die Naturwissenschaften* 11 (1923): 873–875.

Sommerfeld, Arnold. "Zur Quantentheorie der Spektrallinien." *Annalen der Physik* 51 (1916): 125–167.

Sommerfeld, Arnold. "Die Drudesche Dispersionstheorie vom Standpunkte des Bohrschen Modelles und die Konstitution von H2, O2, and N2." *Annalen der Physik* 53 (1917): 497–550.

Sommerfeld, Arnold. "Indische Reiseeindrücke." *Zeitwende* 5 (1919): 289–298.

Sommerfeld, Arnold and W. Kossel. "Auswahlprinzip und verschiebungssatz bei serienspektren." *Verhandlungen der Deutschen Physikalischen Gesellschaft* 21 (1919): 240–259.

Sommerfeld, Arnold. *Atombau und Spektrallinien*. 1st ed. Braunschweig: Vieweg, 1919.

Sonnert, Gerhard. *Einstein and Culture*. Amherst: Humanity Books, 2005.

Sorokin, Pitirim Aleksandrovich. *Social Mobility*. New York and London: Harper and Brothers, 1927.

Sprinker, Michael. *Edward Said: A Critical Reader*. New York: Wiley-Blackwell, 1993.

Stachel, John. "Einstein and Bose." In John Stachel (ed.), *Einstein from 'B' to 'Z'*, 519–538. Boston: Birkhauser, 2002.

Staley, Richard. *Einstein's Generation: The Origins of the Relativity Revolution*. Chicago: University of Chicago Press, 2008.

Staley, Richard. "On the co-creation of classical and modern physics." *Isis* 96, no. 4 (2005): 530–558.

Stanley, M. "Einstein's generation: The origin of the relativity revolution [book review]." *The British Journal for the History of Science* 42 (2009): 470–471.

Steinmann, Ralph M. *Guru-Sisya-Sambandha*. Stuttgart: Steiner-Verlag-Wiesbaden-GmbH, 1986.

Stigler, Stephen. "Stigler's law of eponymy." In *Science and Social Structure. A Festschrift for Robert K. Merton, Transactions of The New York Academy of Sciences*, 147–157, Series II, 39. New York: The New York Academy of Sciences, 1980.

Storer, Norman. *The Sociology of Science: Theoretical and Empirical Investigations*, 286–324. Chicago and London: The University of Chicago Press, 1973.

Stuewer, Roger. *The Compton Effect: Turning Point in Physics*. New York: Science History Publications, 1975.

Subbarayappa, B. V. *In Pursuit of Excellence: A History of the Indian Institute of Science*. Bombay: Tata McGraw Hill Publishing Company Ltd, 1992.

Subrahmanyam, Sanjay. *Is 'Indian Civilization' a Myth?*. Ranikhet: Permanent Black, 2013.

Sur, Abha. "Aesthetics, authority, and control in an Indian laboratory: The Raman- born controversy on lattice dynamics." *Isis* 30, no. 1 (1999): 25–49.

Sur, Abha. "Scientism and social justice: Meghnad Saha's critique of the state of science in India." *Historical Studies in the Physical Sciences* 33 (2002): 87–105.

Sur, Abha. *Dispersed Radiance: Caster, Gender and Modern Science in India*. New Delhi: Navayana, 2011.

Sutton, Geoffrey V. *Science for a Polite Society: Gender, Culture, and the Demonstration of Enlightenment*. Boulder: Westview, 1995.

Swinne, Edgar. Richard Gans: *Hochschullehrer in Deutschland und Argentinien*. Berlin: ERS-Verlag, 1992.

Tagore, Rabindranath. *Viswa-Parichaya (An Introduction to the Universe)*. Calcutta: Viswa-Bharati, 1937.

Tagore, Rabindranath. *Our Universe*, trans. I. Dutt. Bombay: Jaico, 1969.

Tennyson, Jonathan. *Astronomical Spectroscopy: An Introduction to the Atomic and Molecular Physics of Astronomical Spectra*. Singapore: World Scientific, 2011.

Ter Haar, D. *Master of Theory: The Scientific Contributions of H. A. Kramers*. Princeton: Princeton University Press, 1998.

The Shaping of Indian Science: Indian Science Congress Association, Presidential Addresses. Vol 1: 1914–1947. Hyderabad: Universities Press, 2003.

The Twenty-Seventh North American Bengali Conference Journal. Detroit, Michigan. (2007): 77–79.

Uberoi, J. P. S. *The Other Mind of Europe*. Oxford: Oxford University Press, 1984.

Van Vleck, J. H. "The absorption of radiation by multiply periodic orbits, and its relation to the correspondence principle and the Rayeigh-Jeans law. Part I. Some extensions of the correspondence principle." *Physical Review* 24 (1924): 330–346. Reprinted in (Van der Waerden, 1968, pp. 203–222).

Van Vleck, J. H. *Quantum Principles and Line Spectra*. Washington DC: National Research Council Bulletin of the National Research Council 10, Part 4, 1926.

Venkataraman, G. *Journey into Light: Life and Science of C. V Raman*. New Delhi: Penguin books India Ltd., 1994.

Venkataraman, G. *Raman and His Effect*. Hyderabad: Universities Press, 1995.

Venkataraman, G. *Saha and His Formula*. Hyderabad: Universities Press, 1995.

Visvanathan, Shiv. *Organizing for Science: The Making of an Industrial Research Laboratory*. New Delhi: Oxford University Press, 1985.

Vivekananda, Swami. "The Ether." *New York Medical Times* (1895): 58.

Waerden, B. L. van der. *Sources of Quantum Mechanics*. Amsterdam: North Holland Pub. Co., 1967.

Wali, Kameshwar. *Chandra: A Biography of S. Chandrasekhar*. Chicago: University of Chicago Press, 1991.

Warwick, Andrew. "On the role of the Fitzgerald-Lorentz contraction hypothesis in the development of Joseph Larmor's electronic theory of matter." *Archive for the History of Exact Sciences* 43 (1991): 29–91.

Warwick, Andrew. *Masters of Theory: Cambridge and the Rise of Mathematical Physics*. Chicago: University of Chicago Press, 2003.

Washbrook, D. "Orientalism and the postcolonial predicament: Islam and modernities [book review]." *History Worskshop Journal* 38 (1994): 256–258.

Weber, Max. *Essays in Sociology*. New York: Oxford University Press, 1946.

Young, Robert. *Postcolonialism: An Historical Introduction*. Oxford and Malden: Blackwell Publishers; Chennai: T.R. Publications, 2001.

Zachariah, Benjamin. *Nehru*. London: Routledge, 2004.

Zachariah, Benjamin. *Developing India: An Intellectual and Social History, c. 1930–50*. New Delhi: Oxford University Press, 2005.

Zachariah, Benjamin. *Playing the Nation Game: The Ambiguities of Nationalism in India*. New Delhi: Yoda Press, 2011.

Appendix

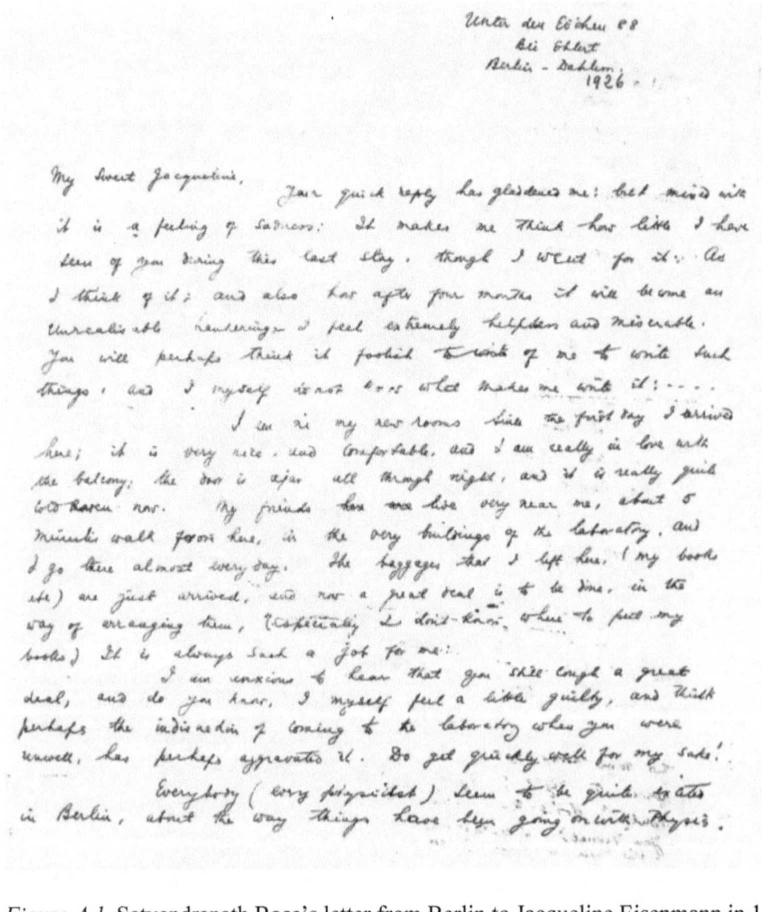

Unter den Eichen 88
bei Ehlert
Berlin - Dahlem.
1926

My Sweet Jacqueline,

Your quick reply has gladdened me: but mixed with it is a feeling of sadness: It makes me think how little I have seen of you during this last stay, though I went for it. As I think of it; and also how after four months it will become an unrealisable hankering, I feel extremely helpless and miserable. You will perhaps think it foolish of me to write such things, and I myself do not know what makes me write it: ....

I am in my new rooms since the first day I arrived here; it is very nice, and comfortable, and I am really in love with the balcony; the door is ajar all through night, and it is really quite cold now. My friends live very near me, about 5 minutes walk from here, in the very buildings of the laboratory, and I go there almost every day. The baggages that I left here, (my books etc.) are just arrived, and now a great deal is to be done, in the way of arranging them, (especially I don't know where to put my books) It is always such a job for me:

I am anxious to hear that you still cough a great deal, and do you know, I myself feel a little guilty, and think perhaps the indiscretion of coming to the laboratory when you were unwell, has perhaps aggravated it. Do get quickly well for my sake!

Everybody (every physicist) seem to be quite excited in Berlin, about the way things have been going on with Physics.

Figure A.1 Satyendranath Bose's letter from Berlin to Jacqueline Eisenmann in 1926.

first on the 28th last Heisenberg spoken in the Colloquium about his theory; then, in the last colloquium, there was a long lecture on the recent hypothesis of the spinning electron. (perhaps you have heard of it) Everybody is quite bewildered and then there is going to be very soon a discussion of Schrödinger's papers. Einstein seems quite excited about it; the other day coming from the Colloquium, we both found him, jumping in the same compartment, where we were, and forthwith he began to talk excitedly about the things we have just heard. He has to admit that it seems a tremendous thing, considering the lot of things which these new theories correlate and explain, but he is very much troubled by the uncomeliness of it all. We were all silent, but he talked almost all the time; unconscious of the interest ~~everybody is taking~~ and wonder that he is exiting in the ~~mind~~ mind of the other passengers.

I have made an honest resolution of working hard during these months, but it is so hard to begin, when once you have given up the habit: though there is here but little else to do.

I am glad that you have liked "the life of Buddha". I only looked at the first few pages; and I was a little afraid that you might not, like it: It would have been such a pity. Apart from all his doctrines, and the myths that have accumulated round him, he is such a noble and outstanding figure; in world history and one whom I adore the most. I wondered when you wanted to read his life, and ~~otherwise~~, never told me before.

It is at after all such a great thing to give you two more books, (books, which I so much like), and have I not even the right of asking you to read what I myself so much like? But I don't know when they will reach you — I have just finished your Proust; I really want to write so much, but it is so late, and ~~this paper is finished~~; more in my next. My best love; Yours; Bors

Figure A.1 (Continued).

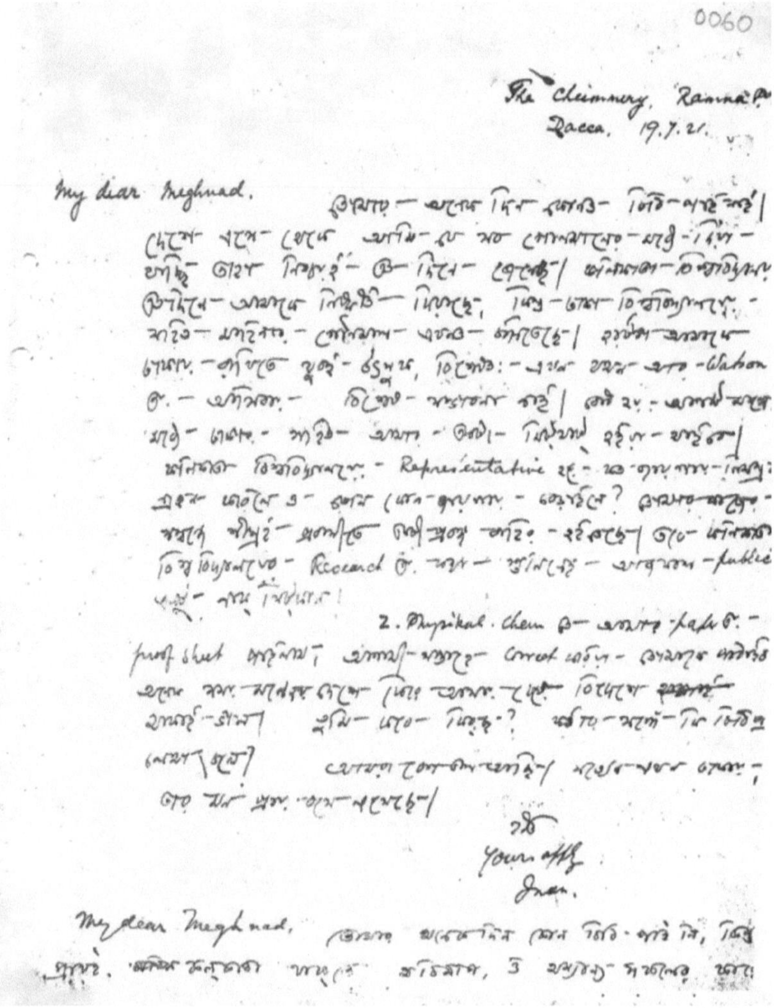

Figure A.2 Indian chemist Jnan Ghosh's letter in Bengali to Saha and Satyendranath Bose's letter in Bengali to Saha.

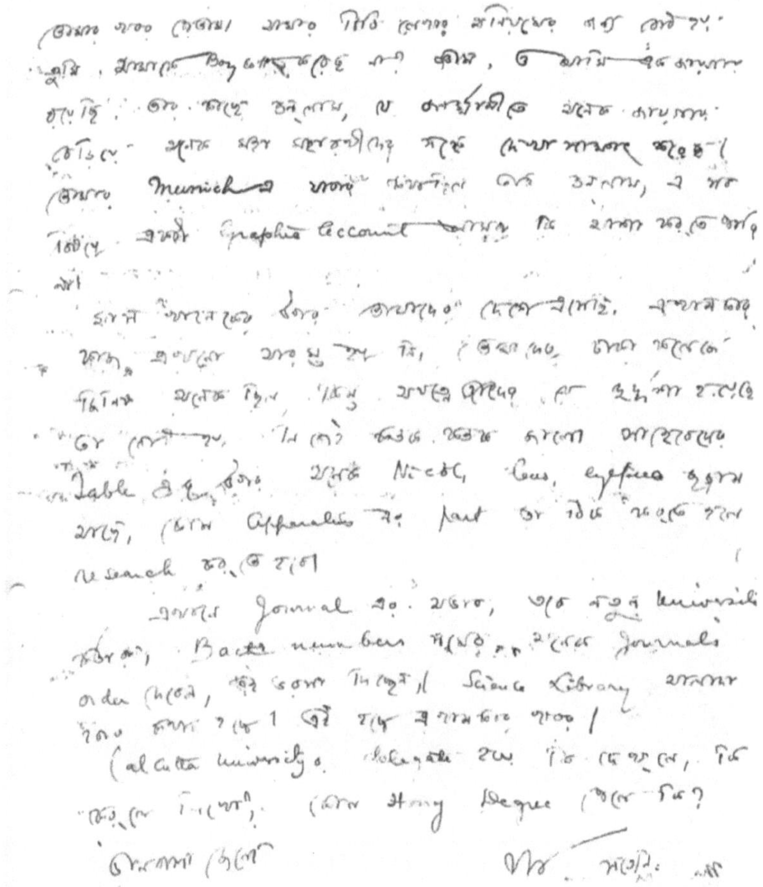

Figure A.2 (Continued).

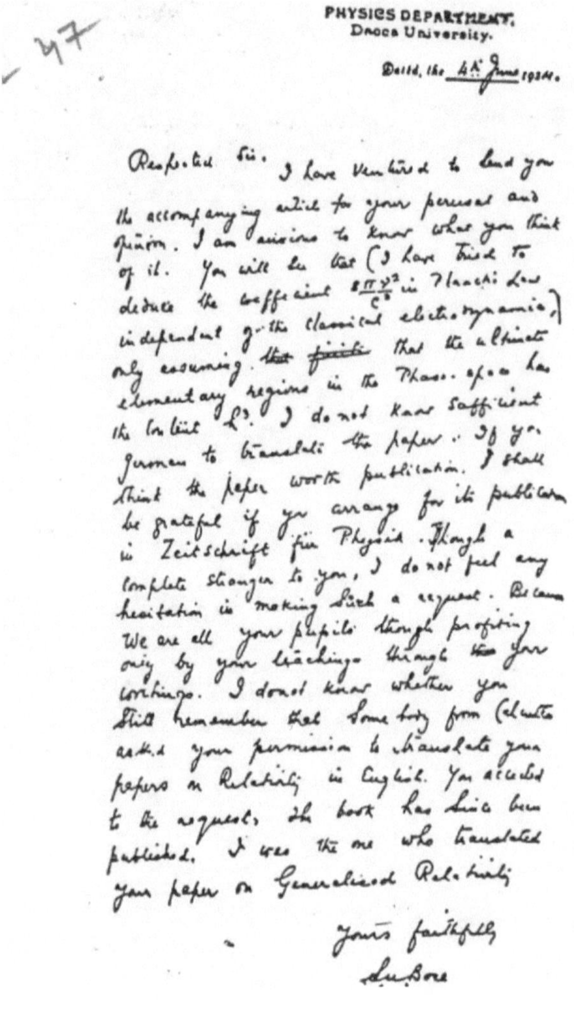

47

PHYSICS DEPARTMENT,
Dacca University.

Dated, the 4th June 1924.

Respected Sir, I have ventured to send you the accompanying article for your perusal and opinion. I am anxious to know what you think of it. You will see that I have tried to deduce the coefficient $\frac{8\pi\nu^2}{c^3}$ in Planck's Law independent of the classical electrodynamics, only assuming that the ultimate elementary regions in the Phase-space has the content h^3. I do not know sufficient German to translate the paper. If you think the paper worth publication, I shall be grateful if you arrange for its publication in Zeitschrift für Physik. Though a complete stranger to you, I do not feel any hesitation in making such a request. Because we are all your pupils though profiting only by your teachings through your writings. I do not know whether you still remember that somebody from Calcutta asked your permission to translate your papers on Relativity in English. You acceded to the request. The book has since been published. I was the one who translated your paper on Generalised Relativity.

Yours faithfully
S.N. Bose

Figure A.3 Satyendranath Bose's letter to Albert Einstein June 4, 1924.

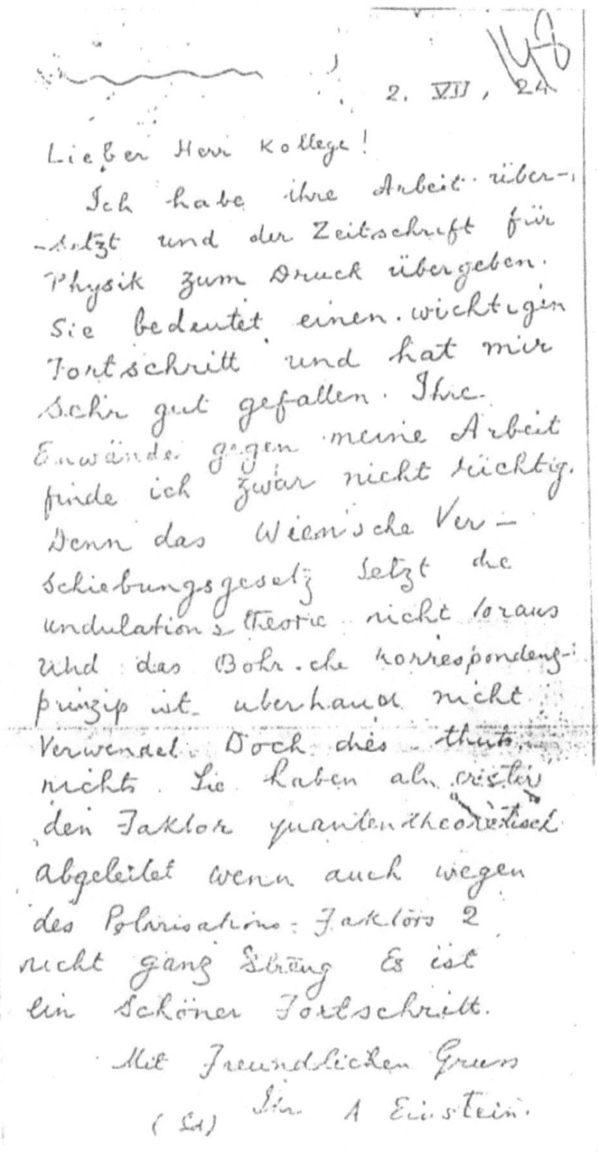

Lieber Herr Kollege!

Ich habe ihre Arbeit über-
setzt und der Zeitschrift für
Physik zum Druck übergeben.
Sie bedeutet einen wichtigen
Fortschritt und hat mir
sehr gut gefallen. Ihre
Einwände gegen meine Arbeit
finde ich zwar nicht richtig.
Denn das Wien'sche Ver-
schiebungsgesetz setzt die
Undulationstheorie nicht voraus
und das Bohr'sche Korrespondenz-
prinzip ist überhaupt nicht
verwendet. Doch dies thut
nichts. Sie haben als erster
den Faktor quantentheoretisch
abgeleitet wenn auch wegen
des Polarisations-Faktors 2
nicht ganz streng. Es ist
ein schöner Fortschritt.

Mit freundlichen Gruss

(24)   Ihr A Einstein.

Figure A.4 Einstein's postcard to Satyendranath Bose July 2, 1924.

2.7.24

Dear Colleague,

    I have translated your work and communicated it to Zeitschrift fuer Physik for publication. It signifies an important step forward and I liked it very much. In fact I find your objections against my work not correct. For the Wien's displacement law does not assume the wave (undulation) theory and the Bohr's correspondence principle is not at all applicable. However, this does not matter. You have derived the factor quantum theoretically for the first time even though the factor 2 due to polarization is not wholly rigorous. It is a beautiful step forward.

    With friendly greeting,

Yours A. Einstein

Figure A.5 Einstein's postcard to Satyendranath Bose July 2, 1924 (translated in English).

17 Rue du Sommerard
Paris V⁔.
27ᵗʰ January '25.

Revered Master.

I received your kind note of 3ʳᵈ Nov. in which you mentioned your objections against the elementary law of Probability. I have been thinking about your objections all along. and so did not answer immediately. It seems to me that there is a way out of the difficulty; and I have written down my ideas in the form of a paper which I am sending under a separate Cover. It seems that the ofpone hypothesis of negative. Einstahlung. stands. which as you have yourself expressed, reflects. the classical behaviour of a resonator in a fluctuating field. But the additional hypothesis of a spontaneous change, independent of the state of the field seems to me not necessary. I have tried to look at the radiation field from a new standpoint, and have sought. to separate the propagation of quantum of energy from the propagation of electro-magnetic influence) I seem to feel vaguely that some such separation is necessary

6

Figure A.6 Satyendranath Bose's third letter to Einstein January 27, 1925.

of Quantum theory – to be brought in line with the Generalised Relativity theory.

The ideas about the radiation-field, which I have ventured to put forward seem to be very much like of what Bohr has recently expressed in Phil. mag 1924. But it is only a guess, as I cannot say honestly to have exactly understood all what he means to say, about his Virtual fields and virtual oscillators.

I am rather anxious to know your opinion about it. I have shown it to Prof Langevin here and he seems to think it interesting, and worth publishing.

I cannot exactly express how grateful I feel for your encouragement, and the interest you have taken in my papers. Your first p. card came at a critical moment, and it has more than any other made this sojourn to Europe possible for me. I am thinking of going to Berlin at the end of this winter, where I hope to have your inestimable help and guidance.

Yours Sincerely
S. N. Bose.

Figure A.6 (Continued).

Berlin, 9 May 1926

Mr. Satyendranath Bose is at present conducting a physical experimental investigation about the refractive index of Roentgen rays in the Kaiser Wilhelm Institute in Berlin-Dahlem. His large and profound knowledge which stretches over the whole of physics as well as the wider territory of chemistry marks Mr. Bose out prominently. His most valuable quality which makes him of inestimable value to a collaborator is his deep and clear insight in the fundamentals of our science, his wealth of fruitful ideas and his capacity to combine theoretically important questions with experimentally feasible tests in an exceedingly happy manner. From the presence of Mr. Bose, our institute derives the greatest benefit all the more as he understands, in a masterly way, how to make the most difficult questions clear through discussion – a quality which seems to make Mr. Bose eminently suitable specially to the profession of teaching.

Dr. Hermann Mark
Kaiser Wilhelm Institute
for Chemistry of Fibrous Materials
16 Fambay Road, Berlin-Dahlem

Figure A.7 Hermann Mark letter of reference for Bose, May 9, 1926.

Paris, the 26th April 1926
10 Rue Vauquelin

.... que Francaise
re' Egalite' FRaternite'
Ecole Municipale de
Physique et de Chimie
Industrielles
Office of Director

    I the undersigned, Paul Langevin, Professor at College de France, Director of Ecole de Physique et Chimie de Paris, certify that Mr. Satyendranath Bose, Reader of Physics at the University of Dacca, has spent a year in Paris in 1924-25 and worked under my direction.

    I have the highest esteem for the personal merit and the works of Mr. Bose whose own researches have been pursued here and whose presence has contributed to increasing the scientific activity of our university. I am particularly happy that the initintive of the University of Dacca has permitted this useful stay of Mr. Bose in Paris.

Sd/

P. Langevin

7

Figure A.8 Paul Langevin letter of reference for Bose, April 26, 1926.

Index